T0189349

Lecture Notes in Networks and Systems

Volume 827

Series Editor

Janusz Kacprzyk ⓘ, Systems Research Institute, Polish Academy of Sciences, Warsaw, Poland

Advisory Editors

Fernando Gomide, Department of Computer Engineering and Automation—DCA, School of Electrical and Computer Engineering—FEEC, University of Campinas—UNICAMP, São Paulo, Brazil

Okyay Kaynak, Department of Electrical and Electronic Engineering, Bogazici University, Istanbul, Türkiye

Derong Liu, Department of Electrical and Computer Engineering, University of Illinois at Chicago, Chicago, USA
  Institute of Automation, Chinese Academy of Sciences, Beijing, China

Witold Pedrycz, Department of Electrical and Computer Engineering, University of Alberta, Alberta, Canada
  Systems Research Institute, Polish Academy of Sciences, Warsaw, Poland

Marios M. Polycarpou, Department of Electrical and Computer Engineering, KIOS Research Center for Intelligent Systems and Networks, University of Cyprus, Nicosia, Cyprus

Imre J. Rudas, Óbuda University, Budapest, Hungary

Jun Wang, Department of Computer Science, City University of Hong Kong, Kowloon, Hong Kong

The series "Lecture Notes in Networks and Systems" publishes the latest developments in Networks and Systems—quickly, informally and with high quality. Original research reported in proceedings and post-proceedings represents the core of LNNS.

Volumes published in LNNS embrace all aspects and subfields of, as well as new challenges in, Networks and Systems.

The series contains proceedings and edited volumes in systems and networks, spanning the areas of Cyber-Physical Systems, Autonomous Systems, Sensor Networks, Control Systems, Energy Systems, Automotive Systems, Biological Systems, Vehicular Networking and Connected Vehicles, Aerospace Systems, Automation, Manufacturing, Smart Grids, Nonlinear Systems, Power Systems, Robotics, Social Systems, Economic Systems and other. Of particular value to both the contributors and the readership are the short publication timeframe and the world-wide distribution and exposure which enable both a wide and rapid dissemination of research output.

The series covers the theory, applications, and perspectives on the state of the art and future developments relevant to systems and networks, decision making, control, complex processes and related areas, as embedded in the fields of interdisciplinary and applied sciences, engineering, computer science, physics, economics, social, and life sciences, as well as the paradigms and methodologies behind them.

Indexed by SCOPUS, INSPEC, WTI Frankfurt eG, zbMATH, SCImago.

All books published in the series are submitted for consideration in Web of Science.

For proposals from Asia please contact Aninda Bose (aninda.bose@springer.com).

Zuzana Dvořáková · Anastasia Kulachinskaya
Editors

Digital Transformation: What is the Impact on Workers Today?

Springer

Editors
Zuzana Dvořáková
Masaryk Institute of Advanced Studies
Czech Technical University in Prague
Prague, Czech Republic

Anastasia Kulachinskaya
Graduate School of Industrial Economics
Peter the Great St. Petersburg Polytechnic
University
Saint Petersburg, Russia

ISSN 2367-3370   ISSN 2367-3389 (electronic)
Lecture Notes in Networks and Systems
ISBN 978-3-031-47693-8   ISBN 978-3-031-47694-5 (eBook)
https://doi.org/10.1007/978-3-031-47694-5

Preface

Dear colleagues,

In this book, we have collected a variety of studies to highlight the use of different technologies in education and acquisition of new digital competencies by graduates, as well as to consider the requirements for a modern worker and his working conditions.

Who is he—a worker of new times? What competencies should he have?
What is a smart workplace and what is the modern labor market like?
How should modern students be taught?

We are sure you will find answers to these and other questions in the book.
Enjoy reading!

Prague, Czech Republic
Saint Petersburg, Russia

Prof. Zuzana Dvořáková
Dr. Anastasia Kulachinskaya

Contents

Human Capital in the Digital Economy: Search for a Perspective Field of Research

Ekaterina Fedorova⑩, Olga Kalinina⑩, Xinmin Peng⑩, and Viktoria Vilken⑩

Abstract The chapter provides a brief overview of the trends and structure of the development of the theory of human capital, taking into account the trends of digitalization. The data of publication activity testifying to the exponential growth of scientific interest in the theory of human capital in the international scientific community are presented. The study was conducted for the period 2012–2022, the main focus was on publications for the period 2018–2022. Based on the research of publications, the structure of the development of the theory of human capital in the digital economy has been developed. The influence of the development of means, tools, methods and possibilities of digitalization on various fields of research included in the general theory of human capital is considered. Problematic and promising areas of research in the development of the theory of human capital are identified and schematically displayed, taking into account the directions of general economic theory and the conditions of the modern economy. Conclusions are drawn about the universal nature of the integration of digital trends into various aspects of theory and practice. The arguments in favor of the development of research for the development of human capital management mechanisms on the part of the state in cooperation with institutions and business are presented.

Keywords Human capital · Digital economy · Digital transformation · Human capital management · Human capital development

E. Fedorova (✉) · O. Kalinina · V. Vilken
Peter the Great St. Petersburg Polytechnic University, St. Petersburg, Russia
e-mail: fedorova_es@spbstu.ru

X. Peng
Ningbo University, Ningbo, China

Z. Dvořáková and A. Kulachinskaya (eds.), *Digital Transformation: What is the Impact on Workers Today?*, Lecture Notes in Networks and Systems 827,
https://doi.org/10.1007/978-3-031-47694-5_1

1   Introduction

The development of human capital in the context of digitalization is an urgent topic of modern economic research, the interest in which in international publications has been stable for more than 10 years. This is clearly evidenced by the graph of publication activity compiled based on the results of the Google Scholar platform data for 2012–2022 (see Fig. 1). At the same time, during this period, there is even a fairly active exponential growth, which clearly indicates an increase in the relevance of research on the theory of human capital at the present stage of research.

As the review of the topics of scientific publications for the same period showed, the field of human capital research is quite highly fragmented. Certain aspects of the theory of human capital are considered from different sides separately, with almost no attempts to integrate them into the general theory. It is difficult to say for what reason this feature of research directions arose, but it can be assumed that the reason lies in the qualitative nature of the concept of human capital itself. The absence of a single definition, which has not yet been formulated in theory, generates a lot of additions to existing definitions dictated by the understanding of human capital, taking into account current trends in economics and the development of general economic theory, as well as related studies. Accordingly, all additions to the definition and structure of human capital naturally cause not only expansion, but complication of the structure of the basic theory. The development of the scientific theory of human capital with an emphasis on digitalization is a trend of the last decade, due to social development and increased interest in the problems of digitalization and its impact on various aspects of scientific directions. At the same time, researchers face difficulties in building relationships between the results of their research with the general theory of human capital due to the lack of unified approaches. As a result, this leads to an even greater general fragmentation in the structure of the theory.

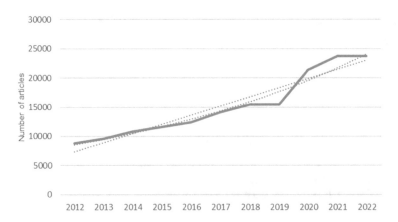

Fig. 1 Schedule of publication activity in the period from 2012 to 2022 on (number of articles containing "human capital" in their title)

Presumably, interest in the development and systematization of the theory of human capital will be stimulated by the development of general economic theory, which increasingly pays attention to the human factor and the systematic management of it in all areas of human activity. In addition, digitalization opens up new opportunities for the collection and analysis of data included in the structure of human capital or related to it.

2   Materials and Methods

The research is based on materials from open sources and official websites. In the process of scientific research, the following nationwide research methods were used: statistical, descriptive, graphic.

The subject of the research is the structure and content of theoretical knowledge at the present stage of studying human capital and its forming factors in the conditions of digitalization.

The methodological basis of the research is theoretical and methodological conclusions and provisions of economic theory in the field of human capital theory.

3   Results

According to the results of the conducted research, the following main areas of knowledge can be identified at the moment, in which the scientific interest of researchers of the theory of human capital in the conditions of digitalization is concentrated: general aspects of the theory of human capital, assessment of human capital and calculation of related indicators, factors of growth and development of human capital, mechanisms and methods of human capital development.

General aspects of the theory of human capital. Despite the absence of a single conceptual apparatus in the theory of human capital in the international scientific community, this issue is not particularly considered in the publications. Researchers rather go to the knowledge of the essence of human capital from the factors that form it and the relationships. Scientists are much more interested in the development of methods for calculating human capital, modeling the processes of its development and calculating investments, rather than the accuracy of formulations. At the same time, it is recognized that inaccurate terminology causes some discomfort in research [1]. It is assumed that an expanded interpretation of human capital will make it possible to form a multilateral state strategy for more effective management, taking into account various aspects of its development. Digitalization also brings its own emphasis to the concept of human capital, since in modern society it is closely related to its formation. And this is of particular interest for further research, since it is necessary to understand what aspects of digitalization should be taken into account in the theory of human capital and how to do it.

Another direction in the development of the theory of human capital can be considered the search for interrelations between the concepts and essence of human capital management and human resource management. In modern conditions, we are witnessing a situation where companies in a number of countries, in conditions of a shortage of qualified personnel in the field of information technology, partially assume the function of forming human capital, providing training and development of actual and future employees. At the same time, there is an inaccuracy of the boundary in science with respect to the concepts and functions of personnel management or human resource management and human capital management. It can be noted that the unifying condition of their relationship is the strategic level of management, and the search for areas of integration is of some scientific interest [2–4]. Thus, research on the formation of human capital in the field of information technology for the development of digitalization of industry will contribute to the development of the theory of human capital in general.

An extensive interest in the development of the theory is the relationship between labor productivity and human capital. It is known that intellectual education is one of the types of economic capital and the level of education determines professional results and the standard of living of people. However, the verification of this theory in recent studies suggests that it needs to be revised and the development of new more complex principles of interconnection [5–7]. For example, it is necessary to investigate exactly how education increases labor productivity or how social status affects education and further the formation of human capital. Individual attempts to find relationships in this area may face imperfection of accounting methods and calculation of individual indicators and even models. One of such barriers to adequate calculation may be the difficulty in determining the educational status of digital industry workers. Often, specialists in the field of information technology, who have one formal level of education, independently improve their qualifications. And if the growth of education through official sources, such as professional development centers and various courses, can be tracked, then independent professional development is not recorded in any way. However, at the same time, it can significantly affect wages, which will indirectly indicate the growth of human capital through the factor of education.

Speaking about the factors of human capital development, attention should be paid to the increased interest of researchers in studying the impact of digitalization on individual sectors of the economy. It is obvious that the introduction of digital technologies into a particular industry indirectly contributes to the growth of human capital. This is confirmed by the results of various studies [8–12]. At the same time, the question of the magnitude of this influence in various industries remains unexplored. The answer to it will help determine the prospects for investment in individual industries and predict the formation of human capital and labor productivity, taking into account this heterogeneity.

Digitalization can have a significant impact on such areas of human capital research that study the impact of people's behavior in choosing a profession, education and behavior in career development. There is an analogy with research methods in marketing, when based on big data about various aspects of people's behavior and their social and psychological characteristics, it is possible to build behavior models and predict choices. This promising area of research creates prerequisites for combining economic, sociological and psychological knowledge to study the features of the formation of human capital [13].

Assessment of human capital and calculation of related indicators. Since the exact structure of human capital is not defined in the scientific community, there is also no unified methodology for calculating it. There is a noticeable tendency to expand the structure of human capital, due to the inclusion of new qualitative indicators related to competence, intelligence, social interaction and the impact of digitalization. The indicated tendency to complicate the calculation of human capital and the inclusion of a large number of indicators in its composition is undoubtedly stimulated by the capabilities of modern information technologies and also by the current scientific trend of searching for non-obvious interdependencies [14]. As for the practical side of research in this direction, there is an obvious interest of the state in obtaining calculation methods that will best reflect the estimated indicators of investment in human capital and the resulting stock of human capital [15, 16]. It is fair to note that spending on medicine and education accounts for a significant share of the budgets of many countries. The need to develop the digital economy also leads to an additional increase in education costs. It is obvious that in order to adequately assess the invested funds and predict returns, new progressive calculation methods are needed, taking into account many aspects of those that were mentioned in the previous paragraph.

In our opinion, the main difficulty in the development and application of evaluation indicators is the factor of constant variability of human capital. Most of the existing methods measure conditional input and output indicators of human capital through the recording of certain events, such as completion of training, obtaining additional qualifications, work experience, or through costs and income. However, all these indicators fix the final state, whereas the formation of human capital can be represented as a constantly changing flow. In the modern world, which is subject to frequent changes, such delays in assessing the state of human capital can lead to the choice of an ineffective strategy.

The second difficulty in assessing human capital is the lack of uniform methods for assessing and taking into account real human competencies, i.e. behavior that collectively either contributes to productivity growth or, conversely, reduces it. Conditionally, we equally evaluate people who have the same level of education and professional experience in calculating human capital, but at the same time we understand that the return on their work can be very different [1]. Previously, there were no methods for such an assessment due to the volume of data and the associated costs. Now digitalization makes it possible to carry out such an assessment, not only to collect data using various sensors and software, but also to process them in real time to build models that reflect the real state of human capital and allow making a more accurate forecast. This is certainly not easy and, in addition, to achieve the goal of collecting such data, it will require work at the level of individual organizations [17], as well as improving methods for evaluating intellectual work.

The factors of growth and development of human capital are determined by its structure. The absence in theory of a unified approach to the formation of the structure of human capital logically stimulates the development of research in the direction of finding the causes that affect it. At the moment, researchers agree that the main factors of growth and development of human capital are income, education, standard of living, health, the economy of the country and the region [14, 18–20]. It should be noted that in the last five years, special attention has been paid to environmental and sustainable development issues in the study of human capital development factors [21–30]. However, all these factors are not homogeneous and independent in themselves. Each of the factors represents systems and institutions in which human capital is one, but not the only result. Moreover, these factors are interdependent and the quality of their interrelations also affects the indicator of human capital, since it combines their combination [6].

Many researchers pay attention to the study of human capital factors in the regions for building models, forecasting development and applying data in the formation of regional strategies [31, 32]. Of course, the possibilities of this field of research are far from being exhausted.

Such a complex structure of the field that forms human capital causes a lot of versatile and quite promising research in this direction. At the moment, it is quite obvious that digitalization is not only a powerful factor for the development of human capital [33], but also provides researchers with appropriate analytical tools. In addition, the digital industries themselves and Industry 4.0 have a significant impact on the formation of human capital [34, 35].

Mechanisms and methods of human capital development. Management of human capital development factors is possible through appropriate mechanisms and methods. As we have already mentioned, digitalization provides more opportunities for research, forecasting, and, as a result, human capital management. In our study, we want to draw attention to the fact that not much attention is paid to these issues in the scientific literature. As a rule, scientists offer various recommendations for the development of human capital directly related to the factors of influence under study, for example, education, medicine or ecology [36–38]. At the same time, we noticed that many of the authors mentioned in this article consider human capital management as a problem and task of organizations, not the state. The role of the state in the development of human capital is practically not mentioned or considered in the research. We are of the opinion that this position of researchers is due to the capitalist way of life of the countries where these studies are conducted [17]. At the same time, we also believe that effective management of human capital is possible only with an economic system that is not aimed at exclusive profit, but at the comprehensive and harmonious development of the state and society.

Digitalization is one of the most important conditions and at the same time factors of human capital development. At the same time, large-scale projects on digitalization of individual industries and areas of activity require large investments, coordination of many institutions, reorientation in the educational environment and other changes. Such projects, including with the involvement of funds from interested investors, are expedient and effective to implement through the state management system. Here we see a significant field for further research.

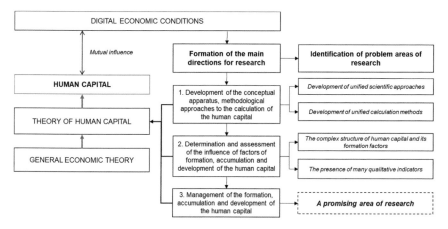

Fig. 2 The structure of the development of the theory of human capital in the digital economy (author's development)

The general process of development of the theory of human capital in the digital economy can be presented structurally in Fig. 2.

According to the specified facts about the state of the theory of human capital in digital transformation, we have now formulated the following distinctive characteristics of it:

a. disparity of scientific principles and approaches;
b. the constant growth of relationships with other scientific fields;
c. the complexity of the tools for conducting research;
d. extensive research potential.

Thus, in the course of the study, it was possible to identify the main structure of the topics of modern research on the theory of human capital, taking into account the trends of digital transformation, and give its brief description. This result can be used to rethink current areas of theory and used as a basis for searching and choosing a narrower research topic.

4 Discussion

The formation of human capital in the digital economy is a complex multifaceted process that is influenced by many factors and conditions. In modern conditions, it becomes obvious that effective management of human capital, by which we mean its sustainable development and accumulation of resulting capital, can be achieved only with close cooperation of government, institutions and private business. We see this as an important area of research that will open up the possibility of managing human capital in accordance with the needs of industries, regions and countries and taking into account their condition.

We see the main difficulty in developing adequate mechanisms of interaction between business and the state, based on serious technical support and means of digitalization, in which information is transmitted to the top in a timely manner, and in response, prompt and targeted measures are followed, allowing to react sensitively to the changes taking place. At the same time, such a cognitive factor as trust in interaction becomes an important issue.

Such a mechanism can be developed taking into account the interests of all participants. However, at the same time, we may encounter a lack of interest in participating in cooperation among small organizations that do not have experience in strategic human resource management and long-term programs for the formation of human capital. But we can also count on the support of large enterprises for which the benefits of cooperation are obvious. Thus, the development of management mechanisms may include a preliminary study of the areas of interest of different parties, the elaboration of their possible strategies taking into account new opportunities, information measures to attract cooperation and, of course, the development of a technical system capable of ensuring this interaction.

5 Conclusion

As a result, we can present several interrelated conclusions.

Human capital studies represent many facets of one phenomenon, each of which is a complex multicomponent environment. For this reason, human capital research will remain quite fragmented in the near future.

In the formation and development of human capital, digitalization acts as the most important factor in its development, as well as an auxiliary management mechanism.

Currently, any of the areas presented in our study can be considered promising due to the many unresolved interdependent scientific issues and practical tasks.

Taking into account the challenges of the modern economy, when many global and country processes are being restructured, the development of effective mechanisms for managing human capital becomes the most important task of researchers.

Acknowledgements The research is partially funded by the Ministry of Science and Higher Education of the Russian Federation under the strategic academic leadership program 'Priority 2030' (Agreement 075-15-2021-1333 dated 30 September 2021).

References

1. Deming, D.: Four facts about human capital. J. Econ. Perspect. **36**(3), 75–102 (2022)
2. Boon, C., Eckardt, R., Lepak, D., et al.: Integrating strategic human capital and strategic human resource management. Int. J. Hum. Resour. Manag. **29**(1), 34–67 (2018)
3. Kucharčíková, A., Mičiak, M., Bartošová, A., et al.: Human capital management and Industry 4.0. In: SHS Web of Conferences, vol. 90, p. 01010. EDP Sciences, China (2021)

4. Kalinina, O., Valebnikova, O.: Human capital management as innovation technologies for municipal organization. In: International Scientific Conference Energy Management of Municipal Transportation Facilities and Transport EMMFT 2017, pp. 1315–1322 (2018)
5. Marginson, S.: Limitations of human capital theory. Stud. High. Educ. **42**(2), 287–301 (2019)
6. Ali, M., Egbetokun, A., Memon, M.: Human capital, social capabilities and economic growth. Environ. Sci. Pollut. Res. **26**(26) (2019)
7. Xiang, C., Yeaple, S., Ali, M., Egbetokun, A., Memon, M.: The production of cognitive and non-cognitive human capital in the global economy. NBER Working Paper No. 24524 April 2018, Revised February 2019, JEL No. F16, F63, F66, I21, I25, I26, O15, O43, O47 (2018)
8. Kuznetsova, I.: The impact of human capital on engineering innovations. Int. J. Emerg. Trends Eng. Res. **8**(2), 333–338 (2020)
9. Zaborovskaia, O., Nadezhina, O., Avduevskaya, E.: The impact of digitalization on the formation of human capital at the regional level. J. Open Innov.: Technol., Mark., Complex. **6**(4), 1–24 (2020)
10. Kichigin, O., Gonin, D.: Human capital as a catalyst for digitalization of regional economy. In: IOP Conference Series: Materials Science and Engineering. IoPscience, Russian Federation (2020)
11. Grigorescu, A., Pelinescu, E., Ion, A.E., et al.: Human capital in digital economy: an empirical analysis of central and Eastern European countries from the European Union. Sustainability (Switzerland) **13**(4), 1–21 (2021)
12. Song, S., Shi, X., Song, G., et al.: Linking digitalization and human capital to shape supply chain integration in omni-channel retailing. Ind. Manag. Data Syst. **121**(11), 2298–2317 (2021)
13. Wright, P.: Rediscovering the "human" in strategic human capital. Hum. Resour. Manag. Rev. **31**(4), 100781 (2021)
14. Angrist, N., Djankov, S., Goldberg, P., et al.: Measuring human capital. SSRN Electron. J. (2019)
15. Abraham, K., Mallat, J.: Measuring human capital. J. Econ. Perspect. **36**(3), 103–129 (2022)
16. Liu, G., Fraumeni, B.: A brief introduction to human capital measures. NBER Working Paper No. 27561, JEL No. J24, O57 (2020)
17. Culpepper, P.T.: Creating Cooperation. Cornell Uneversity Press, San Diego, California, USA (2002)
18. Currie, G.: Child health as human capital. Health Econ. **29**(4), 452–463 (2020)
19. Black, R., Liu, L., Hartwig, F., et al.: Health and development from preconception to 20 years of age and human capital. Lancet **399**(10336), 1730–1740 (2022)
20. Zhou, G., Gong, K., Luo, S., et al.: Inclusive finance, human capital and regional economic growth in China. Sustainability **10**(4), 1194 (2018)
21. Graff, Z.J., Hsiang, S., Neidell, M.: Temperature and human capital in the short and long run. J. Assoc. Environ. Resour. Econ. **5**(1), 77–105 (2018)
22. Yao, Y., Ivanovski, K., Inekwe, J., et al.: Human capital and CO_2 emissions in the long run. Energy Econ. **91**, 104907 (2020)
23. Kim, D., Go, S.: Human capital and environmental sustainability. Sustainability (Switzerland) **12**(11), 4736 (2020)
24. Huang, S., Chien, F., Sadiq, M.: A gateway towards a sustainable environment in emerging countries: the nexus between green energy and human capital. Econ. Res.-Ekon. Istraživanja **35**(1), 4159–4176 (2022)
25. Ahmad, M., Ahmed, Z., Yang, X., et al.: Financial development and environmental degradation: do human capital and institutional quality make a difference? Gondwana Res. **105**, 299–310 (2022)
26. Pata, U., Caglar, A., Kartal, M., et al.: Evaluation of the role of clean energy technologies, human capital, urbanization, and income on the environmental quality in the United States. J. Clean. Prod. **402**, 136802 (2023)
27. Shahbaz, M., Song, M., Ahmad, S., et al.: Does economic growth stimulate energy consumption? The role of human capital and R&D expenditures in China. Energy Econ. **105**, 105662 (2022)

28. Samour, A., Adebayo, T., Agyekum, E., et al.: Insights from BRICS-T economies on the impact of human capital and renewable electricity consumption on environmental quality. Sci. Rep. **402**, 136802 (2023)
29. Ahmed, Z., Wang, Z.: Investigating the impact of human capital on the ecological footprint in India: an empirical analysis. Environ. Sci. Pollut. Res. **26**(26) (2019)
30. Shela, V., Ramayah, T., Noor, H.A.: Human capital and organisational resilience in the context of manufacturing: a systematic literature review. J. Intellect. Cap. **24**(2), 535–559 (2023)
31. Gillman, M.: Steps in industrial development through human capital deepening. Econ. Model. **99**(3) (2021)
32. Zhang, Y., Kumar, S., Huang, X., et al.: Human capital quality and the regional economic growth: evidence from China. J. Asian Econ. **86**, 101593 (2023)
33. Li, H., Zhuge, R., Han, J., et al.: Research on the impact of digital inclusive finance on rural human capital accumulation: a case study of China. Front. Environ. Sci. **10** (2022)
34. Flores, E., Xu, X., Lu, Y.: Human Capital 4.0: a workforce competence typology for Industry 4.0. J. Manuf. Technol. Manag. **31**(4), 687–703 (2020)
35. Kelchevskaya, N.R., Shirinkina, E.V.: Evaluation of digital development of human capital of enterprises. In: 2nd International Conference on Education Science and Social Development, pp. 446–449. Atlantis Press, China (2019)
36. Khalil, S., Shah, S., Khalil, S.: Sustaining work outcomes through human capital sustainability leadership: knowledge sharing behaviour as an underlining mechanism. Leadersh. Org. Dev. J. **42**(7), 1119–1135 (2021)
37. Diaz-Delgado, M., Gil, H., Oltra-Badenes, R., et al.: Detonating factors of collaborative innovation from the human capital management. J. Enterprising Communities: People Places Glob. Econ. **14**(1), 145–160 (2019)
38. Peng, X., Lockett, M., Liu, D., Qi, B.: Building dynamic capability through sequential ambidexterity: a case study of the transformation of a latecomer firm in China. J. Manag. Organ. **28**(3), 1–20 (2022)

Smart Work and Lifelong Learning for Workers' Employability

Kryštof Šulc⬤ and Zuzana Dvořáková⬤

Abstract Smart work and lifelong learning are intertwined concepts that empower individuals to thrive in a dynamic and evolving world. Workers can enhance their employability, productivity, adaptability, and professional growth through training that increases chances of success in both personal and professional spheres. Industrial robotics impacts jobs in small- and medium-sized enterprises (SMEs) as it affects job tasks and the competences required by occupations. Employees must collaborate with robots and use their competences in combination with technology. Robotization can reduce employability in unskilled and semi-skilled jobs like handlers, warehouse workers, and assembly workers. For enhancing their employability, smart work plays a critical role. A necessary pre-condition is that employers invest in training aimed mainly at programming, robot control, data analytics, and cyber security. It can increase productivity, reduce production costs, improve product quality, cut production time, and improve flexibility. SMEs usually have limited financial resources and must carefully consider each investment. So, the rate of return on investment is a critical factor in the decision of industrial SMEs to adopt new technologies and digitize operations regardless of their crucial role in the labor market.

Keywords Smart work · Digitization · Learning · Small and medium-sized enterprises · Robotization · Employee

K. Šulc
Czech Institute of Informatics, Robotics and Cybernetics, Czech Technical University in Prague, Jugoslávských partyzánů 1580/3, 160 00 Prague 6, Czech Republic
e-mail: krystof.sulc@cvut.cz

Z. Dvořáková (✉)
Masaryk Institute of Advanced Studies, Czech Technical University in Prague, Kolejní 2637/2a, 160 00 Prague 6, Czech Republic
e-mail: zuzana.dvorakova@cvut.cz

Z. Dvořáková and A. Kulachinskaya (eds.), *Digital Transformation: What is the Impact on Workers Today?*, Lecture Notes in Networks and Systems 827,
https://doi.org/10.1007/978-3-031-47694-5_2

1 Introduction

Smart work refers to working intelligently and efficiently, utilizing resources, skills, and strategies to achieve desired outcomes effectively and with optimal productivity. On the one side, it involves applying critical thinking, leveraging technology, prioritizing tasks, and managing time effectively to maximize results. On the other one, it brings the intelligent use of technology, tools, and strategies to optimize work processes and achieve better outcomes. It involves leveraging digital tools, automation, collaboration platforms, and data-driven insights to work more efficiently and effectively to simplify tasks, improve collaboration, and make informed decisions [1]. Smart work leads to changes like:

- Automating repetitive and mundane tasks using technologies like robotic process automation and artificial intelligence enables employees to focus on more value-added activities,
- Using digital collaboration tools, project management platforms, and virtual communication channels to facilitate seamless teamwork, irrespective of geographical boundaries,
- Encouraging an agile mindset and flexible work arrangements, e.g., remote work, to adapt to changing business needs and improve work-life balance,
- Promoting a culture of continuous learning and upskilling to stay updated with the latest tools and technologies,
- Utilizing data analytics and business intelligence tools to gather insights, monitor performance, and make data-driven decisions.

Smart work and lifelong learning develop two essential concepts that go hand in hand in the rapidly changing world. The first mentioned emphasizes optimizing processes, leveraging technology, and making informed decisions to maximize productivity and effectiveness. On the other hand, the second one acts as the ongoing process of acquiring knowledge, skills, and attitudes throughout one's life. It involves a proactive attitude towards personal and professional development, recognizing that learning is not limited to formal education but can occur through various channels such as books, online courses, workshops, mentoring, and practical experience [2].

The connection between smart work and lifelong learning lies in their mutual reinforcement and the demands of the evolving labor market. They complement each other. Life-long learning equips individuals with the skills and knowledge to excel in high-value activities. It helps individuals stay abreast of industry trends, technological advancements, and changing market conditions. Automation and technological advancements driven by Industry 4.0 are disrupting changes, so intelligent work and lifelong learning are crucial for futureproofing one's career [3]. By being adaptable, agile, and continuously learning, individuals can navigate through transitions, acquire new skills, and seize emerging opportunities. In the endless learning process, smart work enables the application of new competences efficiently and effectively. New knowledge and skills can support individuals towards employability when adapting to new challenges and remaining competitive in the labor market.

In summary, smart work and lifelong learning demonstrate intertwined concepts that empower individuals to thrive in a dynamic and evolving world. By adopting a strategic approach to work and embracing continuous learning, individuals can enhance their productivity, adaptability, and professional growth, thereby increasing their chances of success in both personal and professional spheres.

2 Methods

Methods applied use the state-of-art literature review and a case study. They cover an analysis of the secondary sources and a narrative case illustrating a calculation of the rate of return. The analysis of bibliographic data identifies trends, monitors knowledge development, and searches for ideas in the field of work digitalization and learning. The focus is on manual workers who are unskilled and semi-skilled in an industrial company. An emerging employment issue is to achieve employability as knowledge and skills undergo substantial changes due to automatization and robotization. However, training means investment on the side of employers and willingness to life-long learning on the side of employees. Critical factors of robotization illustrate a narrative case study about a Czech foundry that belongs to the segment of industrial small- and medium-sized enterprises (SMEs). It calculates what the employer must consider when adopting new technologies and digitizing operations.

The reason for using these methods is that the topic deals with social phenomena that determine the national culture, economic conditions, and attitudes of workers whose behavior influences a national socio-economic system and labor relations. Combining two methods provides a rational ground that gives a realistic view of intelligent work, training, and robotization in industrial SMEs. It helps predict job requirements that can become a perspective for the digitalization of manual jobs.

3 Results

Modern trends and technologies create an excellent opportunity for transforming manufacturing processes to a higher level, especially when using industrial robotization. The flexibility and possibilities to improve existing manufacturing technologies differ depending on the type of enterprise. Large industrial enterprises have more options and mainly act as leaders in adopting industrial robots. Small and medium-sized enterprises generally have limited facilities, like research and development departments, sufficient financial capital, in-depth market research on new technologies, and other parameters compared with large enterprises. However, they provide conditions for applying industrial robotics as a pathway to higher profitability and reduced cost per unit of final product, specifically in large-scale production [4].

3.1 Jobs and Competences

Industrial robotics significantly impacts unskilled jobs in SMEs as it affects both the job tasks and the competences required by occupations. The substantial consequences happen in the industry where robotization can result in more efficient production processes, reduced costs, and increased productivity. However, employees must collaborate with robots and apply their knowledge and skills with technology [5]. Robotization influences the reduced employability predominantly expected in unskilled occupations due to changes in job tasks before and after robotization, as in Table 1.

So, to others. Communicating clearly, both verbally and in writing, fosters positive workers' employability and can build on learning new competences that match with re-designed work processes. Knowledge of automation, programming, logistics, data analysis, repair, and maintenance of robotic systems can guarantee employment in the labor market. Developing communication and teamwork is also suited to key competencies as workers are required to operate effectively with robots and cooperate

Table 1 The impact of industrial robotics on unskilled occupations in SMEs

Occupation	Before robotics	After robotization
Handlers and warehouse workers	They usually perform manual activities such as moving and packing goods, provisioning, and storage	The need for manual handling and moving goods may decrease. Robots can perform these tasks more quickly and efficiently. Competencies in this area could shift to robot management and supervision, equipment maintenance, and logistics processes
Assembly workers	They perform manual assembly activities, folding parts or assembling products	Robots can perform repetitive assembly tasks more quickly and accurately. Workers could shift to supervising robots, programming their activities, and repairing and maintaining them. They could focus on more complex assembly tasks requiring human creativity and expertise
Testing and inspection workers	They specialize in product quality control, perform testing, and ensure compliance with standards	Some product inspection and testing can be automated. Workers could focus on supervising robots, evaluating test results, troubleshooting problems, and analyzing data obtained from robotic systems. They can also be involved in quality improvement and innovation processes
Cleaners and maintenance workers	They care for equipment maintenance, cleaning, and repairs	They can specialize in robot maintenance and repair. They may become specialists in programming, diagnosing, and repairing robotic systems responsible for planning and optimizing equipment maintenance

Source Gášova, M., Gašo, M., Štefánik, A.: Advanced industrial tools of ergonomics based on Industry 4.0 concept. Procedia Engineering, 219–224 (2017). Edited by the authors with team members [6]. Competent workers recognize the value of collaboration and effective communication as they participate in team efforts, contribute ideas, listen to labor relationships, and convey information to colleagues, clients, and stakeholders.

3.2 Employability of Unskilled and Semi-skilled Workers

Employers can promote a worker's employability as a person's ability to gain and maintain competences supporting their employment. It encompasses knowledge, skills, attitudes, and personal characteristics that make an individual attractive to potential employers. Employers seek candidates with technical knowledge and transferable skills in a competitive labor market, mainly soft skills like problem-solving, critical thinking, time, and task management.

For enhancing employability, innovative work plays a critical role. However, it refers to workers' adaptability and continuous learning. They must actively seek opportunities to develop new skills and quickly adapt to changing work environments to stay updated with industry trends. The decisive aspect belongs to technological proficiency. Competent workers must be comfortable with the technology commonly used in their industry, which covers staying updated with relevant software, tools, and digital platforms.

Smart work can enhance the employability of unskilled and semi-skilled workers. However, employers must invest in their on-the-job training and support the development of soft competences. The focus in their lifelong learning can include fostering adaptability, problem-solving skills, time management, collaboration and communication, and technological proficiency. By developing and demonstrating these skills, workers strengthen their competitiveness in the labor market and perspectives of employment.

3.3 Training Transformation Based on Digitalization

Digitalization and industrial robotics radically force the training and development of workers in SMEs. It is essential to provide adequate training to prepare employees for new technologies and enable them to effectively benefit from digitalization [7]. Changes need both education institutions and training programs provided by employers. Higher education institutions (HEIs) must update study programs to reflect modern trends and technologies, like incorporating industrial robotics, process automation, programming, and digital skills into curricula and courses. Employee training in SMEs must aim at programming, robot control, data analytics, and cyber security. Collaboration between SMEs and HEIs can tap into their expertise and design specialized courses that lead to internships, joint projects, or co-financing training programs.

During the COVID-19 pandemic, digitalization revealed the benefits of online and e-learning as practical means of delivering employee training. Online platforms and virtual courses allow employees to study and learn new skills through interactive materials, videos, tests, and discussion forums. Digital learning tools provide flexible and tailored-made courses, enabling to learn without time and space constraints. However, lifelong learning and retraining expect that employees see learning as a continuous process and are motivated to retrain their competences. On the other side, businesses must promote lifelong learning and offer jobs for reskilled workers. In addition to any theoretical training, workers must gain practical experience in using digital technologies and robotics that may transfer to daily work. SMEs can facilitate this practice through projects where trainees work on actual tasks and apply their skills in concrete work situations. SMEs gradually adapt to the digital environment and take advantage of the benefits of digitalization and industrial robotics. Financial support is essential to invest in life-long training for mastering new technologies and workflows [8].

3.4   Digitization and Robotics Rates in SMEs in the EU

Digitalization and industrial robotics considerably impact European Union (EU) SMEs. SMEs play a vital role in the European economy, so the extent of their digitization and robotization give opportunities and challenges. According to studies by the European Commission, a considerable gap exists between large enterprises and SMEs in digitalization. In 2020, only 1% of EU enterprises with at least ten persons employed reached a very high level of digital intensity [9]. In 2018, even 7% of EU enterprises employing at least ten persons used industrial or service robots [10]. Many SMEs rely on traditional practices and need more skills and financial resources to fulfil their digital potential. In robotics, they need to be more engaging in the automation and the use of robotic technologies even though they know that technological innovations cause benefits, like automating repetitive processes, reducing costs, and improving manufacturing processes' accuracy, speed, and quality.

The EU has taken several measures to support the digital transformation of SMEs. The initiatives cover the following:

- Financial support, e.g., Horizon Europe and the European Regional Development Fund (ERDF) offer to fund research, innovation, and digitalization,
- Advice and training, i.e., the EU supports the provision of advice and training to SMEs on digitization and robotics to raise awareness of digital technologies,
- Networking and know-how sharing, i.e., the EU supports the creation of networks and platforms for SMEs to share their experiences, best practices, and know-how in the field of digital transformation,
- Cooperation with educational institutions by promoting cooperation between SMEs and educational institutions such as HEIs, technical schools, and vocational colleges,

- Support access to infrastructure to improve SMEs' access to digital infrastructure and broadband internet.

Although the EU is taking initiatives towards the digital transformation of SMEs, some barriers threshold changes. Critical obstacles comprise limited funding, insufficient expertise and skills among employees, slowly pathing digitization of public services and problematical access to digital infrastructure in rural and remote areas [10].

Digitalization and robotics can boost the competitiveness of SMEs in the EU. SMEs can develop and contribute to sustainable economic growth and innovations based on the EU financial support, training, and networking cooperation. In Europe, robot installations were up 24% to 84,302 new units in 2021, with steady demand from the automotive industry and demand from the general industry up by 51% [10].

3.5 Potential of Digitization for Employment in SMEs

Industrial robotics and digitalization have great potential for innovations in SMEs and can lead to changes in various job roles [11, 12]. Illustrating the impact, for example:

- Handlers and warehouse workers: Robots can be programmed to move, pack, and store goods quickly and accurately. Handlers and warehouse workers can then supervise robots, manage warehouse systems, optimize logistics processes, and manage inventory.
- Assembly workers: Automation of assembly processes allows streamlining production as industrial robots can perform repetitive assembly tasks faster and more accurately. Assembly workers can move on to more complex tasks that require human creativity and expertise, such as quality control, debugging and repairing robotic systems, innovating manufacturing processes, and developing new products.
- Testing and inspection workers: Industrial robots and sensors can inspect products and test processes with high accuracy and speed. Test and inspection workers can focus on evaluating test results, analyzing robot data, quality management, and process improvement.
- Cleaners and maintenance workers: Robots can perform routine maintenance, diagnostics, and equipment repair. Cleaners and maintainers can be responsible for programming and supervising robots, scheduling maintenance, managing sensors, and monitoring machine performance. They may also specialize in advanced technologies such as remote diagnostics and predictive maintenance.

In summary, robotics and digitalization in industrial SMEs incorporate increased productivity, lower production costs, improved product quality, reduced production time, and increased flexibility. Workers develop new skills in programming, robot control, data analysis, and process optimization [13]. A pre-condition is to provide training for adapting to new technologies and opportunities to work on advanced tasks related to robotics and digitalization.

4 Discussion

Entering the era of industrial robotics does not only mean a revolution in large industrial enterprises but also medium-sized enterprises such as foundries. One of the vital transformation aspects is the automation of unskilled jobs. A typical case demonstrates foundry workers who work in occupations requiring reskilling if foundries use industrial robots to streamline the production process.

A typical molder characterizes a worker (usually a man) in a foundry specializing in casting metals. The main job is to prepare and operate foundry equipment, such as foundries and molds, which transform the molten metal into the desired shape. A molder works with different metals such as iron, aluminum, or steel and ensures the melt's correct temperature, quantity, and quality during the casting process. The molder is responsible for preparing the foundry molds, filling the melt into the molds, removing defects and bubbles in the product, and checking the quality of the finished parts. In addition, he must be able to set up and operate foundry equipment such as furnaces and casting machines correctly and observe appropriate safety precautions. He must have an excellent technical and mechanical understanding, the ability to operate various tools and measuring equipment, and work cooperatively with team members. The molder follows precise procedures and standards to ensure high-quality end products. In the robotization context of the foundry process, the molder work may undergo changes where some manual tasks may be replaced by automated systems and robots, which increases efficiency and repeatability and reduces the risk of human error. However, foundry workers need training to adapt to new technologies and practices.

An illustration of work changes provides a molder position in a low-pressure aluminum foundry in Czechia when comparing the current job task with replacement by an automatic arm. The casting machines are usually semi-automatic, which can ensure the process of casting aluminum into the mold by themselves. However, it is no longer possible to ensure the process of removing the casting automatically. So, the casting removal process is handled by the caster, which transfers the casting into a standard iron crate at a predetermined location with eye inspection. It is the only task of the molder. Automating the process would be easy, leading to the replacement of the whole process with an automatic arm, i.e., a collaborative robot. The cost of modifying the robot's point of operation would be zero, as there would be no need to install protective cages. The desirability of introducing automation is expressed by a rate of return (RoR). For example, based on the input parameters [14, 15]:

- Robotic arm:
  - Purchase cost including additional elements for the robot 50,000 EUR,
  - Maximum power input 1 kWh,

- Installation cost 63,500 EUR,
- Market price per kW 6 €/kWh,
- Number of shifts per day 3 × 8 h,
- Number of working days per year 2023 is 250.

- Employee

  - Gross salary EUR 1,800 (CSU, average wages in the industry for the 4th quarter of 2022),
  - Total social and health insurance on gross wages in Czechia is 33.8%,
  - Number of shifts per day 3 × 8 h,
  - Number of working days in 2023 is 250.

Calculation:

Average number of hours per month

$$h_m = (d_r/m_r) * h_w = (250/12) * 24 = 500 \text{ h/m} \tag{1}$$

h_m average number of working hours per month.
d_r number of working days per year 2023.
m_r number of months per year.
h_w number of hours worked per day in 3 shifts.

Cost function for a collaborative robot

$$\begin{aligned} f(x_1) &= FC_1 + FC_2 + (P * p_e * h_m) * m \\ &= 50000 + 63500 + (1 * 6 * 500) * m \\ &= 113500 + 3000 * m \end{aligned} \tag{2}$$

FC_1 purchase price of collaborative robot.
FC_2 commissioning and installation cost.
P power consumption of the collaborative robot.
p_e average electricity price.
h_m average number of working hours per month.
m number of months of full operation (variable).

Cost function for three-shift operation

$$f(x_2) = (S + (1 + I) * W_s) * m = (1800 + (1 + 0.334) * 3) * m = \\ = 7203.6 * m \tag{3}$$

S average gross wage per worker.
I social security and health insurance contributions for the employer.
W_s number of shifts per working day.

m number of months of full operation (variable).

Determine the payback period in months for the investment in the collaborative robot:

$$f(x_1) = f(x_2)$$
$$FC + VC_1 = VC_2$$
$$113500 + 3000 * m = 7203.6 * m \qquad (4)$$
$$m = FC / (VC_2 - VC_1) = 113500 / (7203.6 - 3000) \cong 27$$

FC total fixed cost of the collaborative robot.
VC_1 average variable cost per collaborative robot per month.
VC_2 average variable cost per employee per month for a three-shift operation.
m number of months until the payback period of the collaborative robot investment.

According to the simplified calculation of the payback period for the investment in the collaborative robot, the entire investment can recover within 27 months of commissioning compared to a three-shift operation. It is necessary to consider that employees have holidays and are on sick leave. Both circumstances also affect the payback period in favor of the collaborative robot.

5   Conclusion

The rate of return on investment is a critical factor in the decision of industrial SMEs to adopt new technologies and digitize operations. The RoR assessment influences investment decisions [16] as SMEs usually have limited financial resources and must carefully consider each investment. The RoR is an indicator that helps assess whether an investment in new technologies and digitalization can yield a sufficient financial return compared to the costs. A reasonable RoR is essential to maintain SMEs' financial stability and competitiveness.

Factors influencing the RoR cover:

- Investment costs: The high upfront costs of introducing new technologies and digitalization can significantly affect the rate of return. SMEs need to consider whether these costs are justified given the expected benefits and reduction in operating costs in the long term.
- Expected benefits: SMEs need to assess the expected benefits of investing in new technologies and digitization. These include improvements in efficiency, speed, quality of production, reduction of errors, and increased competitiveness. The expected benefits should outweigh the cost of the investment and give SMEs a long-term competitive advantage.

- Payback period: The time it takes to pay for the investment and to generate a positive financial return is also an essential factor in assessing the RoR. SMEs must consider whether they can invest in projects with extended payback periods that may involve higher risks.
- Financial stability and availability of finance: SMEs' financial stability and availability of finance have a significant impact on investment decisions. If funds are limited, SMEs may choose to postpone investments or seek alternative sources of finance.

Further, it is necessary to consider aspects like the following:

- Promoting and stimulating investment: The EU and individual Member States are taking measures to support and stimulate investment in SMEs. It includes the provision of finance through various grants and subsidies, soft loans, and tax breaks that reduce financial barriers and make an investment in new technologies and digitalization more attractive for SMEs.
- Risk assessment: SMEs must carefully assess the risks associated with investments in new technologies and digitalization. It includes identifying technological, operational, personnel, and security risks. The risk assessment enables measures to minimize and manage risks to improve the expected RoR.
- Monitoring and evaluation: SMEs must regularly monitor and evaluate the results of investments in new technologies and digitalization to adjust, when necessary, in approach and strategy towards maximizing the RoR.

The RoR represents an essential indicator for deciding whether to invest in new technologies and digitalization in SMEs. They must carefully assess costs, expected benefits, payback period, and financial stability before investing [17]. With support and incentives from the EU and Member States, they are better placed to overcome financial barriers and increase the RoR on their investments. Monitoring and evaluating the results of investments support optimizing performance and ensuring their long-term competitiveness and employment.

References

1. Raguseo, E., Gastaldi, L., Neirotti, P.: Smart work supporting employees' flexibility through ICT, HR practices and office layout. Evid.-Based HRM: Glob. Forum Empir. Sch. **4**(3), 240–256 (2016)
2. Pacher, C., Woschank, M., Zunk, B.M.: The role of competence profiles in Industry 5.0-related vocational education and training: exemplary development of a competence profile for industrial logistics engineering education. Appl. Sci.-Basel **13**(5), 3280 (2023)
3. Bercovici, E.G., Bercovici, A.: Israeli labor market and the fourth industrial revolution. Amfiteatru Econ. **21**(SI13), 884–895 (2019)
4. Fraga-Lamas, P., et al.: A review on industrial augmented reality systems for the Industry 4.0 shipyard. IEEE Access 13358–13375 (2018)
5. Bartoš, M., et al.: An overview of robot applications in automotive industry. Transp. Res. Procedia 837–844 (2021)

6. Gášova, M., Gašo, M., Štefánik, A.: Advanced industrial tools of ergonomics based on Industry 4.0 concept. Procedia Eng. 219–224 (2017)
7. Habibi, F., Zabardast, M.: Digitalization, education and economic growth: a comparative analysis of Middle East and OECD countries. Technol. Soc. 101370 (2020)
8. Hartong, S.: The transformation of state monitoring systems in Germany and the US: relating the datafication and digitalization of education to the global education industry. In: Researching the Global Education Industry: Commodification, the Market and Business Involvement, pp. 157–180 (2019)
9. Eurostat: How digitalised are EU's enterprises? https://ec.europa.eu/eurostat/en/web/products-eurostat-news/-/ddn-20211029-1. Last accessed 01 July 2023
10. Eurostat: 25% of large enterprises in the EU use robots. https://ec.europa.eu/eurostat/web/products-eurostat-news/-/ddn-20190121-1. Last accessed 01 July 2023
11. Schaupp, E., Abele, E., Metternich, J.: Potentials of digitalization in tool management. Procedia CIRP 144–149 (2017)
12. Deuse, J., et al.: Systematic combination of Lean Management with digitalization to improve production systems on the example of Jidoka 4.0. Int. J. Eng. Bus. Manag. **12** (2020)
13. World Robotics report. https://ec.europa.eu/newsroom/rtd/items/771175/en. Last accessed 01 July 2023
14. ČSÚ: Average wages—4. quarter of 2022. https://www.czso.cz/csu/czso/ari/average-wages-4-quarter-of-2022. Last accessed 01 July 2023
15. Eurostat: Domestic producer prices—energy. https://ec.europa.eu/eurostat/databrowser/view/TEIIS030/default/table?lang=en. Last accessed 01 July 2023
16. Cullinane, S., et al.: Improving sustainability through digitalisation in reverse logistics. In: Digitalization in Maritime and Sustainable Logistics: City Logistics, Port Logistics and Sustainable Supply Chain Management in the Digital Age. Proceedings of the Hamburg International Conference of Logistics (HICL), pp. 185–196 (2017)
17. Alsufyani, N., Gill, A.: Digitalisation performance assessment: a systematic review. Technol. Soc. **68** (2022)

Study on the Quality of Professional and Digital Competences of Master's Students in the Context of Power Engineering and Electrical Engineering Labour Market

Svetlana Nikolaevna Valeeva⑩**, Julya Sergeevna Valeeva**⑩**, and Viktor Vladimirovich Maksimov**

Abstract The present-day labour market sets higher requirements to the qualification of specialists in the field of electrical power engineering and electrical engineering, which is due to the rapid technological development in the electric power market in the digital economy context. The study argues that the high quality of training of master's students in this field is a key factor in their successful career in this industry. The aim of the chapter is firstly, to analyse the quality of master's students training under the program of electric power engineering from the university's perspective, in the needs of the present-day labour market context and secondly, to determine the factors affecting the efficiency of training employees in the energy industry. The chapter used the method of interviewing, which allowed us to conduct a comprehensive analysis across 4 universities selected. As a result of the study, the authors proposed recommendations on improving the quality of higher education in terms of mastering professional and digital competencies, such as updating the curriculum, introducing new technologies, organising practical classes and increasing the number of job opportunities on the labour market for alumni of the electric power engineering master's degree programmes.

Keywords Modernization of education · Profile · Digital competencies · Masters · Energy · Modern labor market

S. N. Valeeva (✉) · J. S. . Valeeva · V. V. Maksimov
Kazan State Power Engineering University, Kazan, Russia
e-mail: esp_snvaleeva@mail.ru

Z. Dvořáková and A. Kulachinskaya (eds.), *Digital Transformation: What is the Impact on Workers Today?*, Lecture Notes in Networks and Systems 827,
https://doi.org/10.1007/978-3-031-47694-5_3

1 Introduction

The matter of quality of specialists' training in the terms of the needs of the present-day labour market is currently extremely relevant both for universities and employers. This global problem requires finding solutions to the following questions: How is the process of training future specialists carried out? How effective and qualitative is the existing training? What professional and digital electric power engineering competences are formed in the course of training of master's students of electric power engineering programme? To what extent do the competences of graduates meet the requirements of professional standards for electric power engineering specialities? What knowledge, skills, and competences, are essential for a successful career in the power industry? Let's consider these questions through the examples of master's degree programmes for the electric power sector at universities in Russia, the CIS and Europe.

The choice of master's programmes in the field of electric power engineering is determined by the fact that in the circumstances of changing conjuncture and transformation of the world energy market it is expedient to consider the quality of educational services for the energy industry. This is an important factor in the replenishment of the managerial staff, as well as in providing the labour market with highly qualified personnel who will have the necessary professional and digital competencies to meet the needs of the industry under study.

The main purpose of the study is to consider theoretical provisions of the content component of education modernisation. Modernisation of education is presented in this study in terms of improving the system of qualitative mastering of professional and digital competences in electric power engineering field at the university, their systematisation; conducting appropriate analytics and developing recommendations. The developed recommendations will allow to generate management decisions and implement measures in the scientific and educational process.

The relevance of this study is conditioned by the expediency of studying theoretical provisions, by the importance of substantiating arguments about the need to improve the system of development of professional and digital competences of master's degree students in the field of electrical power engineering and electrical engineering, because at the moment this issue has not been fully studied. The existing works on the subject do not give a complete picture of the mechanism of education modernisation specifically for the tasks of the energy industry labour market.

It is important to point out that the quality of mastering professional and digital competences can be viewed from several perspectives. Firstly, from the perspective of the university by means of questionnaires, interviewing the teaching staff and administrative management personnel of the university. Secondly, it is possible to consider it from the students' perspective, on the basis of analysing the data obtained as a result of testing, preparation of written works by students. We have limited the scope of our research to analysis from the perspective of professional, digital competences and assessment of university teachers.

2 Literature Review

The analysis of literature sources allowed us to identify a number of factors affecting the quality of training of electric power engineering specialists. In our opinion, important factors were identified in their research by Lucas et al. [1]. The authors identified factors that hinder the development of the RES (Renewable Energy Sources) sector in correlation with the replenishment of human resources, namely:

– shortage of specialists in the field;
– mismatch between the proposed education system and the demand of the sector;
– mismatch between educational programmes for specialists' training and the actual needs of the labour market.

According to expert projections, by 2040 developed European countries will face a severe shortage of specialists for the adaptation and expansion of RES technologies. The authors make suggestions for improving the process of education and training in the field of RES, analysing training courses on the basis of IRELP data.

It can be assumed that the approaches proposed by the authors will lead to an effective solution of a rather complicated situation. Nevertheless, the issue of training modern specialists to meet the requirements of high-tech production should be considered in all aspects.

It should be noted that the rapid development of the digital world and the resulting increase in the need for highly qualified specialists require the development of new tools, approaches and mechanisms for the quality of training of future specialists.

In Russia there is a document that regulates the list of competences of specialists required for employment in production. It is called the professional standard [2]. It describes in detail the labour actions, necessary skills, necessary knowledge that a specialist should possess depending on the qualification.

When analysing this document, the need for its adjustment was identified, given that the digitalisation of the energy sector is developing very rapidly, and high-tech equipment requires the availability of competent professionals with a narrow specialisation.

It was found that cybersecurity specialists are a particular problem, and they are also in short supply at the moment. Since cybersecurity is both a technical and a business risk, Russian companies are concerned about the competences of university graduates.

Georgiadou et al. [3] also addressed the issue of cybersecurity from the perspective that the energy sector is highly vulnerable to cyberattacks due to the complex cyber infrastructure around the world.

When looking at the European labour market for cybersecurity professionals, Blažič [4] suggests that a new learning ecosystem is being created to bridge the gap between the growth of cybersecurity and the growth of advanced technologies, but that industry education needs innovation and renewal.

It is obvious that it is necessary to integrate new topics into the educational programmes for training specialists in this area, supported by practical training, which will certainly serve as a response to the social and economic needs of the labour market.

It is very important to define and introduce into educational programmes a clearer set of competences, and to pay special attention to professional competences, to include more practice in the programmes, and thus to implement in the programmes a practical approach to the training of specialists.

In our opinion, a very interesting opinion on the studied problematic from the perspective of the concept of training industrial workers in the learning factory environment is presented by Büth et al. [5]. The concept is based on energy efficiency and digitalisation trainings in learning enterprises, which significantly increase the efficiency of technology use, respectively increase the level of professional and digital competences of employees.

Regarding the correlation between the educational programmes of universities and real market opportunities, Aljohani et al. [6]. The opinion of these authors is favourable to us, because a clearly structured educational pathway will allow to form a set of necessary professional competences.

The assignment of a mentor of industrial practice to a master's student from the first year of study will allow to build a clear trajectory of professional growth.

The project of innovative higher education programmes of the University of Žilina reviewed by Lusková and Buganová [7] aims to improve the quality of higher education by developing innovative forms, attractive teaching materials and streamlining curricula with a focus on the needs of the labour market and knowledge society. The process of modernisation of university education is considered through the prism of existing cases and attempts of educational structures to reorient themselves towards the new paradigm. It can be assumed that the creation of a new educational programme will be based on the triad of soft, self and hard skills, which will allow to develop mechanisms for the transformation of the programme with the introduction of a set of competencies necessary for a modern specialist.

The issue of transformation of educational programmes is now more relevant than ever. Ruiz-Rivas et al. [8] believe that the need for updated educational programmes is due to the paradigm shift taking place in the energy sector itself. This paradigm represents a significant increase in the use of renewable energy and the need to radically address sustainability issues.

The authors have identified three reasons which, in their opinion, contribute to the transformation of the educational programmes of technical universities:

– digitalisation of the energy sector (digital equipment, digital power grids, digital substations, etc.);
– the need to include sustainable development topics in engineering programmes;
– training of specialists with design skills, knowledge of maintenance and operation of energy systems in the digital economy.

Certainly, all three reasons identified by the authors can influence the process of educational programmes transformation in technical universities. This will allow to develop a mechanism for training qualified specialists.

It should be noted that the process of master's degree training in European universities is often based on the use of innovative software systems. The attendance of students remains a topical issue for all times. The authors Petchamé et al. [9] conducted a study of students and teachers of the master's programme (2021–2022) taught in a hybrid virtual format using pocket Bipolar Laddering tool based on the Smart Classroom system.

The authors' task was to obtain information about the hybrid virtual format and to complete a SWOT analysis in order to decide whether this format should be used in the future, but the main research question was the issue of master's programme processing.

As a result of this study, it was found that students appreciate this format as it allows them to follow the master's programme live and synchronously from abroad during their absence from class.

Certainly, the use of different learning formats is necessary as it stimulates students and keeps them interested in mastering knowledge, skills and abilities.

However, the definition of value orientations and incentives for students in the process of selecting bachelor's degree programmes and admission of master's degree candidates who have shown an inclination to professional activity, as noted in their research Tarasyeva et al. [10] are the most important issues in the quality of training of prospective specialists.

The employability of graduates has become a key factor for universities in Australia and the UK. This problematic is considered in the study by Clarke [11]. Since many universities in Europe now include internships, job placement and international training in their programmes in order to improve the employment prospects of graduates, the authors proposed 6 key dimensions collected into a framework. This framework includes six key dimensions: human capital, social capital, individual attributes, individual behaviour, perceived employability and labour market factors to explain the conceptualisation of graduate employability.

One example of the conceptual framework of graduate employability in Tanzania is the study of Mgaiwa [12]. The authors identified 4 aspects in the employability framework such as: developing effective university-industry partnerships, aligning university education with the country's development plans, regular reviews of university curricula and strengthening quality assurance systems.

When considering this issue, we found that in almost all European universities there is a need to transform educational programmes in order to implement the competences demanded by the modern labour market.

The next important aspect of this study is the consideration of digital competences.

Akimov et al. [13] have outlined a four-dimensional model for measuring competences, which includes assessment of: (1) basic knowledge and skills that contribute to solving everyday tasks, and first of all assessment of information and computer literacy; (2) creativity and critical thinking skills; (3) initiative and self-regulation skills, flexibility and adaptability; (4) the skill of personal cognitive strategy formation formed in the process of meta-learning.

In our opinion, this model can contribute to the measurement of students' digital competences to a greater extent.

A current issue concerning the readiness of Italian universities to implement digital skills and competences in educational programmes is considered in the work of Spada et al. [14]. The authors analysed the most demanded competences in employment. As a result, it was revealed that the process of digitalisation has influenced both the change in competencies demanded by the labour market and their concentration in the digital economy.

The Industrial Revolution (Industry 4.0) is transforming industries all over the world, and the requirements for education systems are changing accordingly. Goldin et al. [15] consider a new educational paradigm of universities through the introduction of new didactic concepts, as well as the use of digital teaching tools in educational programmes that will enable the formation of professional and digital competencies. The authors developed a reference architecture for the integrated and effective use of digital tools in Education 4.0.

The above-mentioned arguments about the modernisation of education, predetermine the necessity to study the quality of competences mastering, as well as the content of the main profile and digital competences for the electric power sector.

Thus, the theoretical basis for the argumentation of the transformation of the educational process is presented, and the competences to be analysed in relation to the profile universities are highlighted.

3   Methods

In the period from May to June in 2023, interviewing of respondents was conducted in order to determine the level and quality of training received by master's students of electric power engineering major in the conditions of education digitalisation. About 12 universities in Russia and CIS countries (Kazan, St. Petersburg, Kazakhstan, Tajikistan, Kyrgyzstan, etc.) participated as respondents.

The range of questions proposed to the respondents for the interview covered all relevant issues in our opinion and is presented below:

1. What in your opinion includes the concept of "digital competences" and "competences of end-to-end technologies" in the educational process and in energy companies?
2. How do you assess the level of your digital competences from 1 to 10?
3. What in your opinion are digital technologies and what are end-to-end technologies?
4. What digital and end-to-end technologies would you like to learn or improve to teach more effectively?
5. How do you feel about the digitalisation of education in general and what benefits can it bring to the training of electrical engineering undergraduates?
6. What do you think are the skills and competences that students in the electric power industry need to master in order to be successful in a digitalised environment?

7. In your opinion, what new professions and specialisations, and consequently new competences may appear in the electric power industry in the conditions of digitalisation? Will the digitalisation of the electric power industry entail the transformation of competences of master's students in the electric power industry?
8. What digital tools and technologies do you use to create and deliver lectures and seminars, as well as to assess students' performance?
9. Do you have new educational programmes jointly with employers taking into account the use of digital and end-to-end technologies?
10. Do you involve practitioners to teach digital competences to students on the basis of the department directly at enterprises?
11. What professional development courses have you taken in this area?

The methodology of this study consisted of several steps.

Before conducting the interview, the main competences were considered taking into account the analysis of literature sources and the state standard for master's programme 13.04.02 "Electric power engineering and electrical engineering" presented by 4 universities. This allowed to form a list of basic profile and numerical competences of master's students of this direction.

At the first stage the respondents had to evaluate the mastering of profile and digital competences of master students of the direction 13.04.02 "Electrical power engineering and electrical engineering" on a 10-point scale.

At the second stage, digital tools and software were classified and systematised according to the criteria of application practice in the activities of electric power companies.

Within the framework of the third stage, a factor analysis of the quality of mastering the profile and digital competences of master's students of the direction 13.04.02 "Electric power engineering and electrical engineering" was carried out. The universities that most effectively organise the educational and research process of master's students were identified.

At the fourth stage the main recommendations for modernisation of educational and scientific processes for master's students of electric power engineering major are proposed.

4 Results

As a result of interviewing the teaching staff of the master's programme of the direction 13.04.02 "Electric power engineering and electrical engineering" of the presented universities, we have identified the profile competences, such as:

1. management of research results in the field of electric power systems, networks, power transmission, their modes, stability and reliability;
2. application of methods of scientific and technical information collection and analysis of research results in the field of professional activity;

3. application of modelling and optimisation methods allowing to predict properties and behaviour of objects in the field of electric power systems, networks, power transmission, their modes, stability and reliability;
4. use of specialised software in conducting research in the field of professional activity.

Digital competences, are presented as follows:

1. Ability to apply advanced theoretical and practical knowledge that is at the forefront of science and technology in the field of professional activity.
2. Ability to professionally operate modern equipment and devices.
3. Readiness to use modern and advanced computer and information technologies.
4. Readiness to use modern achievements of science and advanced technologies in research.

Respondents' assessment of the mastering of professional competences allowed to obtain the results shown in Fig. 1. The following universities are shown below:

– Kazan State Power Engineering University (KSPEU)
– Kyrgyz State Technical University named after I. Razzakov (KSTU)
– Kazakh Agro-Technical Research University (KATRU)
– L.N. Gumilyov Eurasian National University (ENU).

The expert evaluation showed that the mastering of the presented competences takes place in all universities, but it is most effectively reflected in the training of master's degree holders in electric power engineering major of two universities (KSPEU, ENU) (Fig. 2).

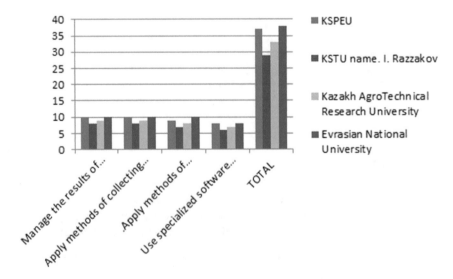

Fig. 1 Ranking of development of professional competences in terms of 4 universities

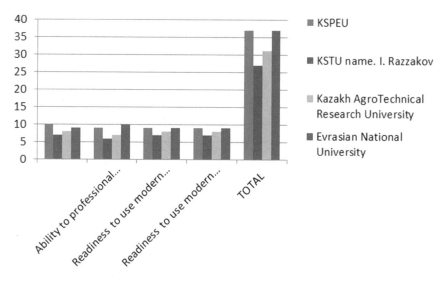

Fig. 2 Rating of digital competence development in terms of 4 universities

The assessment of digital competences shows that not all competences are mastered at a high level during the process of master's degree training programme 13.04.02 "Electric power engineering and electrical engineering".

The study of literature sources, analytics and specifics of electric power enterprises' activities allowed us to identify digital tools, which are classified and presented in Table 1.

According to the results of the assessment of the use of digital tools in the curricula of master's degree programmes 13.04.02 "Power Engineering and Electrical Engineering", we can conclude that functional digital tools are actively used by only two universities in their educational programmes, and the use of sector-specific and operational digital tools is not yet fully implemented in all the universities represented.

The results of the interviews not only revealed positive trends in the process of educating students in the study programme, but also identified a number of problematic issues that need to be addressed. We carried out a factor analysis of important aspects considered in the development of professional and digital competences in master's programmes (Table 2).

Based on the results of the comprehensive analysis, we have developed recommendations that we propose as part of measures to ensure the implementation of digital competences in the educational process with a practical approach to specific enterprises on the side of partners:

– Development of special new disciplines that are based on the mastering of digital technologies and solutions that allow to organise at the enterprises of the energy sector the digitalisation of processes of the main areas of activity: digitisation of

Table 1 Digital tools used in master's degree programmes "Electrical engineering and electrical equipment of enterprises" in the context of 4 universities

Digital tools classification	Digital tools' names	Universities			
		KSPEU	KSTU named after I. Razzakov	KATRU	ENU
		Points out of 10			
1. Sector-specific digital tools	ANSYS programme, Simulink, Statistica SolidWorks и FlowVision Matlab, MathCad, Classic, Maple V Femlab, FlexPDE, FEMM, ELCUT	8	6	5	9
2. Operational digital tools	Bitrix 24, Trello, Jira, Notion, amoCRM	6	5	5	7
3. Functional digital tools	Microsoft Visual Basic 2008; Delphi 7.0; Turbo Pascal; Visual C++ 6.0; Borland C/C++; SKADA System; labVIEW; Zulu, PSCAD programme Consol Multifizics	7	6	6	9

operational activities and application of digital solutions for specific functional tasks.

Designing professional development and retraining courses for students as an additional tool for developing other digital and professional competences that are not defined in the curriculum.

– Dividing and forming a group of students with achievements in creating individual educational trajectories for the development of digital and professional competences.
– External state and public organisations for students to implement the established plan at the university.

In addition, the proposal to organise an individual educational trajectory for masters, which will deepen the knowledge of the discipline, expand applied professional areas of knowledge, as well as focus more on specific segments of the labour market, is relevant.

The next important direction, in our opinion, is the creation of basic departments on the territory of specialised industry organisations, as a rule, industrial enterprises provide laboratory facilities, the opportunities for laboratory workshops and control measurements for the enterprise. This certainly affects the extension of the field of professional competences.

Table 2 Factors influencing the quality of mastering professional and digital competencies through educational programmes of master's degree in electric power engineering major

Positive		Negative	
Internal	External	Internal	External
1. Availability of partner enterprises that provide demo versions of digital programmes for their implementation in the educational process of master's degree students	1. Attracting regional and federal grants for the preparation and creation of new educational programmes using digital competencies and implementation in the curricula of the teaching staff	1. Lack of or low level of digital competencies among teachers to integrate digital tools into educational programmes	1. Lack of access to the acquisition of foreign software
2. Organisation of professional development courses for teaching staff	2. Conducting professional development courses by production mentors	2. Lack of understanding of the future prospects for the use of purchased expensive equipment (in the opinion of teachers)	2. Exodus of programmers who can work as software developers for the energy sector
3. Development of joint master's degree programmes in electric power engineering with employers	3. State subsidy of projects for software development in the energy sector	3. Lack of co-operation among energy enterprises on joint development of curricula	3. Low salaries of teachers at state universities
4. Availability of production mentors, assigned to master's students	4. Energy companies open basic faculties to expand core competences for master's students		

5 Discussion and Conclusion

The analysis of works on the subjects of this study revealed the need to develop a framework for the training of highly qualified specialists for the energy sector, demanded by the labour market through the assessment of competencies. The need to introduce new pedagogical approaches, methods and tools, development and introduction of innovations into existing educational programmes is, in our opinion, an urgent problem facing both Russian and European universities.

We considered the process of modernisation of education in universities through the existing cases from the point of view of the importance of transition of universities to a new educational paradigm. At the same time, diagnostics and statistics of the process of master's students' training will be carried out, subsequently new mechanisms will be developed to facilitate the application of innovative solutions in practice.

Within the bounds of the theoretical provisions of the chapter, arguments are formulated that universities should provide master's students of energy engineering not only with theoretical knowledge and skills, but also with practical experience in energy companies and projects, strive to update educational programmes and use modern educational technologies, provide master's students of energy engineering with modern knowledge and skills.

The scientific significance of this study is the generalisation and systematisation of existing digital solutions and software according to the criterion of the scope of their application at the enterprises of the electric power industry. The results of the summary analytics on the quality of mastering competences, which was carried out on the basis of the expert approach by the point method, made it possible to identify through factor analysis the positive and negative sides in the organisation of education, as well as to develop management solutions for their improvement. The proposed measures are aimed at managerial innovations, at the introduction of pedagogical and educational innovations.

Within the framework of the theoretical block of this study, the works of scientists, who in their published studies have determined and substantiated the expediency and necessity of modernising the master's programmes and improving the quality of mastering the professional and digital competences of master's degree holders of the electric power profile, have been studied. In particular, the works highlighted some elements that were important from our point of view to update and take into account in the practical study.

The following questions were answered in the course of the research: What digital competences have already been implemented in educational programmes and how to assess their application within the practical block of the study based on the interview method. A survey of the teaching staff was conducted in the context of four universities, which made it possible to assess the quality of mastering professional and digital competences in the represented universities.

The interview results were used as a basis for the development of universal proposals for the educational system of universities, which can be used to improve the quality of mastering professional and digital competences of master's students in the electric power engineering major.

References

1. Lucas, H., Pinnington, S., Cabeza, L.F.: Education and training gaps in the renewable energy sector. Sol. Energy **173**, 449–455 (2018). ISSN 0038-092X. https://doi.org/10.1016/j.solener.2018.07.061
2. Electricity. Professional standards. https://fgosvo.ru/docs/index/2/2022

3. Georgiadou, A., Michalitsi-Psarrou, A., Askounis, D.: A security awareness and competency evaluation in the energy sector. Comput. Secur. **129**, 103199 (2023). ISSN 0167-4048. https://doi.org/10.1016/j.cose.2023.103199

4. Blažič, B.J.: The cybersecurity labour shortage in Europe: moving to a new concept for education and training. Technol. Soc. **67**, 101769 (2021). ISSN 0160-791X. https://doi.org/10.1016/j.techsoc.2021.101769

5. Büth, L., Blume, S., Posselt, G., Herrmann, C.: Training concept for and with digitalization in learning factories: an energy efficiency training case. Procedia Manuf. **23**, 171–176 (2018). ISSN 2351-9789. https://doi.org/10.1016/j.promfg.2018.04.012

6. Aljohani, N.R., Aslam, A., Khadidos, A.O., Hassan, S.-U.: Bridging the skill gap between the acquired university curriculum and the requirements of the job market: a data-driven analysis of scientific literature. J. Innov. Knowl. **7**(3) (2022). https://doi.org/10.1016/j.jik.2022.100190

7. Lusková, M., Buganová, K.: Creation and innovation of study programmes with emphasis on the needs of labour market and knowledge society. Procedia - Soc. Behav. Sci. **106**, 739–745 (2013). ISSN 1877-0428. https://doi.org/10.1016/j.sbspro.2013.12.085

8. Ruiz-Rivas, U., Martinez-Crespo, J., Venegas, M., Chinchilla-Sanchez, M.: Energy engineering curricula for sustainable development, considering underserved areas. J. Clean. Prod. **258**, 120960 (2020). ISSN 0959-6526. https://doi.org/10.1016/j.jclepro.2020.120960

9. Petchamé, J., Iriondo, I., Korres, O., Paños-Castro, J.: Digital transformation in higher education: a qualitative evaluative study of a hybrid virtual format using a smart classroom system. Heliyon **9**(6) (2023). ISSN 2405-8440. https://doi.org/10.1016/j.heliyon.2023.e16675

10. Tarasyeva, T.V., Agarkov, G.A., Tarasyev, A.A., Koksharov, V.A.: Modeling the choice of an optimal educational trajectory in the conditions of digital transformation of the economy. IFAC-PapersOnLine **55**(16), 382–387 (2022). ISSN 2405-8963. https://doi.org/10.1016/j.ifacol.2022.09.054

11. Clarke, M.: Rethinking graduate employability: the role of capital, individual attributes and context. Stud. High. Educ. **43**(9), 1–15 (2017). https://doi.org/10.1080/03075079.2017.1294152

12. Mgaiwa, S.: Promoting graduate employment: fostering graduate employability: rethinking Tanzania's university practices. Education **11** (2021). https://doi.org/10.1177/21582440211006709

13. Akimov, N., Kurmanov, N., Uskelenova, A., Aidargaliyeva, N., Mukhiyayeva, D., Rakhimova, S., Raimbekov, B., Utegenova, Z.: Components of education 4.0 in open innovation competence frameworks: systematic review. J. Open Innov.: Technol., Mark., Complex. **9**(3), 100037 (2023). ISSN 2199-8531. https://doi.org/10.1016/j.joitmc.2023.100037

14. Spada, I., Chiarello, F., Barandoni, S., Ruggi, G., Martini, A., Fantoni, G.: Are universities ready to deliver digital skills and competences? A text mining-based case study of marketing courses in Italy. Technol. Forecast. Soc. Chang. **182** (2022). ISSN 0040-1625. https://doi.org/10.1016/j.techfore.2022.121869

15. Goldin, T., Rauch, E., Pacher, C., Woschank, M.: Reference architecture for an integrated and synergetic use of digital tools in education 4.0. Procedia Comput. Sci. **200**, 407–417 (2022). ISSN 1877-0509. https://doi.org/10.1016/j.procs.2022.01.239

Digitalization of Functional Management: The View of Employers and Employees

Alina Kankovskaya and Anna Teslya

Abstract The digital revolution, which began at the end of the twentieth century, significantly accelerated the pace during the covid-19 pandemic and after it. Digitalization covers new spheres of public life, new generations are involved in digital processes in society and business. At the same time, the rate of digitalization varies at enterprises of varying sizes and industries. A person has different levels of competencies as an employee and as a member of the public, respectively, he may be differently involved in digital processes at the enterprise and in society. The study of the impact of digitalization on the work of managers is therefore becoming relevant. The problems of the formation of digital competencies of employees are increasingly raised in scientific publications, however, they are considered separately, and the problem of the development of the competencies of employers and top management has been little studied. The aim of the work is to study the readiness of Russian leaders and top managers of enterprises, on the one hand, and employees, on the other hand, to implement digital platforms and, in the future, digital business models. For these purposes, the authors of the study conducted a survey of managers of companies from various industries and business areas, plus a survey of the personnel reserve (final-year students). The chapter examines the attitude of management and employees to the introduction of electronic document management, taking into account the desired and current level. The direct relationship between the attitude of the head of the enterprise, the level of motivation of the middle management and the degree of digitalization of business processes at the enterprise is revealed.

Keywords Digitalization · Electronic document management · Business · Communication · Digital competencies

A. Kankovskaya (✉) · A. Teslya
Peter the Great St. Petersburg Polytechnic University, Polytechnicheskaya, 29, 195251 St. Petersburg, Russia
e-mail: kankowska_alina@mail.ru

1 Introduction

Digitalization is a natural stage of the information revolution and its evolutionary development can be noted over the past two decades. The COVID-19 pandemic has given a new impetus to digitalization processes—the introduction of digital products, processes and competencies in society and business has significantly accelerated.

As a result, it has become important to investigate the readiness of society, government, infrastructure, workers and employers to participate in digital processes. The focus of scientific research has naturally shifted from solving technical problems and studying the main capabilities of new tools to a comprehensive study of social issues of digitalization. At the same time, the interaction of employers and employees is of particular importance, since both professional and private factors and competencies are manifested in this process.

The social impact of digitalization on the employees of the enterprise and the need to take into account their digital competencies was noted by Meyer [1]. The digital competencies of employees have been studied by many researchers [2–5], but the main objective of these studies is to study the knowledge, skills and abilities of workers, and not their readiness to participate in a digital business model.

In the development of research on the impact of digitalization on personnel, Nedyalkova et al. [6] and Haipeter [7] study the legal aspects of protecting staff from the costs and problems of digitalization. Some researchers, such as Herron [8], Melzer and Dievald [9] pay attention to the problems of bullying and oppression by control from the side of the managers. Such a contrast between employees and employers, in our opinion, incorrectly narrows the problem, while in reality the employer is under the yoke of digitalization no less than the employee.

Walter [10] discusses the impact of digitalization on spirituality in the workplace and the opposite, spiritual factors on the dynamics of digitalization processes. His proposed model, which combines psychology, digitalization, workplace and spirituality, is very interesting and useful for the formation of management decisions. However, this model is limited only by subjective factors, which makes it difficult for managers to use it. The development of a management solution should be based not only on psychological comfort and spirituality, but also on the professional skills of the employee.

Malik et al. [11] studied the digital competencies of employees and staff stress in a digital environment. Based on the analysis of workers of different skill levels and from different industries, they identified the need to develop various programs for professional development and psychological adaptation to the digital business model. This area of research seems to be very relevant and it is desirable to supplement it with a study of employers, their readiness and attitude towards digitalization.

The problem of the impact of digitalization on personnel is indeed a priority, but it must be remembered that the management process includes not only an object, but also a subject of management. And thus, it is important to study the impact of digitalization on employers and, in particular, top management. Eksili [12] and Juhász et al. [13] point out that digitalization is an objectively existing external factor that should be taken into account by both employers and employees. This approach does not take into account the problem of adaptation of both employees and managers to business digitalization, the long-term and multidimensional nature of this process.

The impact of digitalization on management routine has been studied in a large number of papers. Wilkesmann warns against the increase of routine, the predominance of protocols over creative search [14], while Turja et al. [15] investigated the relationship of technological factors, technological well-being of the business and the physical well-being of employees. Warning et al. [16] identifies different types of positions of employees in the enterprise, depending on the possibility of digitalization of the labor process. We agree with the conclusion that it is necessary to educate the creative competencies of employees in the context of the growing pace of digitalization. In this regard, the work of Baesu and Bejinaru [17] is interesting, where the concept of a digital leader is introduced and its main characteristics are shown.

Amankwah-Amoah et al. [18] rightly notes that the Covid-19 pandemic has acted as a catalyst for digitalization processes in society and business. Tartarin et al. [19] explores the changing business models in the Netherlands and Romania due to the combined impact of digitalization and the pandemic. Hamburg [20] dedicates her work to SMEs, rightly considering them the least protected from the risks of digitalization and a pandemic.

However, exploring the transformation of business models, the authors consider employers and senior management either as a kind of constant, or as a kind of element hostile to employees. We can talk about the insufficient research of the social position of employers and top management in relation to digitalization, their readiness to participate in digital processes.

This draws attention to the fact that among Russian and foreign studies, insufficient attention is paid to the issues of assessing the relationship between the competencies of employers and employees. It is necessary to study the view of the problem from both sides.

We set the following tasks:

- Identify and systematize the preferences of senior and middle managers regarding the current level of business digitalization.
- Develop an interview questionnaire aimed at identifying the preferences of senior and middle managers regarding the current and preferred level of business digitalization.
- Select the target group for the interview, covering a wide range of industries and areas of activity. Determine the sample size for the interview.
- Organize the interview process for senior and middle managers.
- Analyze the results obtained and identify the preferences of managers and employees for the implementation and development of digital document management in enterprises.
- Assess whether there are differences in attitudes towards the digitalization of individual business processes among middle and top managers, and identify their reasons.

- Formulate recommendations regarding the organization of business digitalization at the enterprise, taking into account the identified factors.

2   Materials and Methods

The study included two sub-studies—employers and employees.

To study the readiness of employers for the digitalization of functional management, it is proposed to use the interview method, since this method allows for a deep analysis of causal relationships. Employers have different backgrounds and experiences, and the interview allows you to identify and analyze them more effectively than the questionnaire.

The following steps must be taken to achieve the goal of the studying the readiness of employers for the digitalization of business processes in the enterprise:

1. Determining the subject and object of the interview, setting tasks, choosing the type of interview. The subject of the interview is the readiness of senior and middle-level managers for business digitalization. The object of the interview is top and middle-level managers of enterprises of various professional, gender and age groups.
   Research objectives include:
   identifying the attitude of managers to electronic document management;
   finding attitudes towards online communication with superiors and subordinates;
   revealing the willingness of leaders to change.
   The tasks set determined the conduct of an individual semi-standardized interview. At the moment, the interview was conducted once, but can be implemented on a regular basis.
2. Designing an interview, including drawing up an approximate list of questions, developing instructions for the interviewer, choosing the method and means of recording and processing the results.
   The recommended research method is hidden sound recording, which allows you to save all the information received from the respondent without any psychological discomfort.
3. Pilot interview.
4. Clarification and correction of the interview program, editing questions, error analysis.
5. Drawing up a final list of questions,
6. Conducting an interview.
7. Analysis of the results.

It has been empirically proven that a sufficient sample size for interviews for socio-economic purposes is 20–30 respondents [21]. The undertaken study included 26 people and the condition of conformity with the sample design was observed.

To study the readiness of employees to participate in digital business processes of organizations, it is proposed to use the questionnaire method.

A similar algorithm was used to study the opinions of employees

1. Definition of the subject and object of the interview, setting tasks, choosing the type of interview. The topic of the questionnaire is the willingness of employees to digitalize business. The subject of the interview are graduate students with working experience and students of short-term training programs.
   The objectives of the study include:
   to identify the attitude of employees to electronic document management;
   clarifying the attitude to online communication with managers and colleagues;
   identification of readiness for change.
   The tasks set determined the conduct of a face-to-face selective survey. This survey can be conducted on a regular basis.
2. Planning of the questionnaire, including the preparation of a questionnaire, the development of instructions for the interviewer, the choice of the method of processing the results.
3. Pilot survey.
4. Clarification and correction of the questionnaire, editing of questions, error analysis.
5. Preparation of the final version of the questionnaire.
6. Conducting a survey.
7. Analysis of the results obtained.

The focus of the research is aimed at students, because, firstly, it is the personnel reserve of business, and secondly, students represent various branches of economics and types of enterprises and organizations, which allows covering various spheres of the economy with minimal sample sizes. The age and gender structure of the sample corresponds to the characteristics of the general population. The total number of respondents is 73 people, which is enough for a study using the method of small sample size [22].

3   Materials and Methods

3.1   Employers

The interview was attended by top and middle level managers of international companies operating in various sectors of the economy—logistics, production, IT, construction, finance etc. The interviews were attended by senior and mid-level managers of international companies operating in various sectors of the economy—logistics, manufacturing, IT, construction and finance. The respondents are heads of Russian companies operating abroad at various levels of foreign economic activity and international business, but mainly in the export format. Representatives of branches and other subsidiaries of foreign TNCs did not participate in the survey.

The interviews were conducted with equal representation of senior and middle management, with male respondents dominating (65% of respondents).

Managers with different work experience were also interviewed, while the group with 10–20 years of managerial experience prevails both in the sample (58 and 27%) and in the companies studied.

In general, all respondents have a positive attitude towards the growing use of ICT technologies in the business processes of their enterprises, focusing on electronic workflow and the possibility of videoconferencing with colleagues. However, dissatisfaction with the digitalization of business processes is also present.

It should be noted that representatives of the highest level of management, regardless of length of work and management experience, were unanimous in their assessments of the document flow—everyone noted a reduction in the time for working with documents. In particular, the CEO of a construction company notes that the electronic document circulation in his enterprise is close to 100%, and it is very convenient both in intra-company and inter-company relations. Similar responses were received from other managers involved in strategic management of their enterprises.

It should be noted that the senior functional managers noted a large discrepancy between the existing and desired level of electronic document management. The estimation of time gains is also lower. The general opinion of the respondents in this group is that the digitalization of business processes is objectively determined, it is impossible to avoid it and it is not necessary to escape it, so the management of the company should focus on digitalization, pool disparate digital resources into a unified system and accelerate electronic workflow.

Middle managers demonstrate the variety of assessments and a lower level of satisfaction with electronic document management. The main reasons common to all enterprises and industries are the duplication of electronic and paper documents, technically deficient systems. At the same time, middle managers estimate the current level of digitalization of the document circulation as 40–60%, and the desired, provided a systematic approach, as 60–80%. But some of them doubt the expediency of more digitalization—they do not see the gain in time in the electronic workflow and note the difficulties in mastering new systems.

Thus, there are three approaches to the implementation and development of digital document management in the enterprise.

1. The head of the enterprise positively assesses the possibilities of electronic document management and implements it on a systematic basis—this increases the efficiency of managerial routine activities, contributes to the growth of the effectiveness of the work of strategic, line and functional managers at the enterprise. Managers subordinate to such a leader successfully improve their ICT skills, they are ready for further growth in the digitalization of business processes and are highly motivated.
2. The head of the enterprise introduces electronic document management from case to case, not seeing it as an effective tool. In this case, information systems are fragmented, top and middle-level managers use the capabilities of ICT and are ready to develop them depending on their personal readiness for the growth

of digitalization and on the quality of the software products they use. Overall, the implementation of ICT is successful but slow. The motivation for the growth of digital competencies among managers of all levels in such an enterprise is low.

3. The head of the enterprise does not see any benefit in the growth of electronic document management and the use of any software products. In this case, the fragmentation of digital products is even higher (as a rule, they are involved in purely functional units—warehouse, accounting), and most managers are not motivated to develop their digital competencies and maintain the traditional approach to organizing management activities.

A separate section of the interview was devoted to interpersonal communication.

It should be noted that the desire to maximize interpersonal communication with colleagues, subordinates and partners at the level of 60–80% prevails among senior managers. Two reasons were given during the interview:

1. High workload and intensity of work, in which the translation of meetings and business communication online significantly frees up time;
2. Personal preferences for reducing face-to-face communication. In this case, it seems most likely that the desire to reduce online communication is also due to psychological overload when working under high uncertainty.

Mid-level managers want more personal communication with colleagues, superior managers, subordinates and partners—40–60% of all communication time. They find personal contact to be more effective. In assessing interpersonal business communication, they emphasize that personal communication, unlike document management, requires personal contact, communication, interaction, and give informal results that cannot be achieved through online communication.

The preferred format of various types of business communication—meetings, negotiations, interviews, briefings, trainings, professional development—is also different for different groups of managers:

1. Senior managers, mainly business leaders, who tend to maximize the online format of all types of communications.
2. Senior executives, including both business leaders and line and functional managers, who strive for 100% online meetings, a high degree of online negotiations and interviews. Routine communications in the field of advanced training and coaching should be clearly online, negotiations and interviews change form depending on the stage (at the beginning of establishing communication a face-to-face meeting is preferable, then online). Clearly, a large number of standard managerial tasks, tight time frames, and a large number of personal interactions dictate this choice.
3. Senior managers face a high degree of uncertainty of the tasks to be solved (primarily strategic management). Non-standard problems usually require personal contacts and interaction. Therefore, meetings and negotiations are preferred offline.
   Middle managers in general, as already mentioned, tend to prefer face-to-face interaction. They prefer routine meetings, professional training and coaching

in online format In relation to other types of business communication, their opinion is conditioned, like that of senior managers, by the ratio of routine and non-standard tasks.

4. Middle managers engaged in solving non-standard tasks prefer the average level of online communication during negotiations and interviews. As well as senior managers they note the need to establish face-to-face contact before communication can be transferred to an online format.

5. Middle-level managers engaged in solving standard tasks, prefer offline communication and talk about the need to avoid the routine, which increases with online communication.

In the course of the study the relations with foreign partners were not identified as a separate group of issues, but the interview revealed that in most cases the interaction with foreign partners is similar, Only the process of initial networking becomes longer.

Summarizing the results of the interview and evaluating the attitude of the management of companies of different levels and fields of activity to the process of digitalization and electronic document management, we can draw the following finding.

The general attitude of the respondents regarding the use of ICT is positive, but the respondents can be divided into three groups: "actively implementing", "implementing as needed" and "resisting". The highest level of management belongs only to the first two groups, The functional management group notes a large gap between the level of electronic document management and the desired one. The general opinion of the respondents in this group is that the digitalization of business processes is objectively determined, it is impossible and it is not necessary to avoid it, so the management of the company should focus on digitalization, link disparate digital resources into a single system and speed up electronic document management. The middle management group generally showed less commitment to the digitalization process.

Thus, the task of developing recommendations for the inclusion of staff in ICT skills training is to work with mid-level managers, novice managers and future managers from personnel reserve.

3.2 Employees

The questionnaire offered to current and future employees included sections on preferred types of interaction with colleagues and managers and on digital competencies.

When communicating with colleagues, 84.6% of respondents prefer to use a messenger, 80.8%—face-to-face contacts and 53.8% use a phone for this.

When it comes to interaction with customers, the gap in ratings for different types of connections is reduced: e-mail (53.8%), face-to-face contacts (65.4%), telephone

(73.1%), and messengers (73.1%)—probably the reason is that the respondents do not choose the type of interaction with the buyer on their own.

When interacting with managers, employees prefer personal contacts (80.8%). As with interaction with clients, the second place is taken by the type of communication that is determined by the business process of the organization—telephone (71.2%), messengers (69.9%).

Figure 1 shows respondents' assessments of their level of proficiency in various ICT skills (1 is the minimum score, 5 is the maximum score). It is noticeable that the survey participants tend to give themselves maximum marks only in matters of communication, otherwise, the respondents rate their knowledge of software use and data analysis as good, the biggest problem for respondents is development. This type of competence ownership can be called one of the most difficult, since it involves a creative approach and work at the level of a technical process.

Respondents rate the attractiveness of professions based on these skills as high as possible, the spread of values here is insignificant, however, the most popular area is communication-related activities (45 people), and the least popular is digital content development (31 people). For most parameters, the proportion of low ratings of all the areas under consideration is minimal. All directions are of great interest to respondents, while the majority of respondents (61.5%) want to solve creative tasks, 26.9%—standard and only 11.5%-unique. The last three questions in the block relate to the consideration of one process from a different angle—document flow. 57.7% would prefer to receive 80–100% of documents in electronic form, while only 44% are ready to read them in this form, but already 76.9% want to use exclusively electronic document flow when sending. The respondents have difficulty and obvious reluctance sending paper documents, this practice has mostly disappeared from everyday life.

Based on the analysis of the results of the survey of employees, the following conclusions can be drawn:

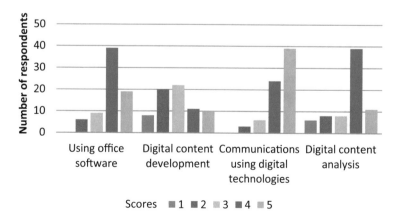

Fig. 1 Assessment of the level of possession of ICT skills

1. If there is a free choice, respondents choose messengers or face-to-face meetings to interact with colleagues. Preference is given to balancing the use of ICT in everyday life.
2. Respondents rate their ICT competencies highly, giving priority to communication skills, while the worst expectations regarding their competencies are related to the development of digital products.
3. All the presented categories of digital professions are of interest to the respondents, the greatest interest is related to the spheres of communication, the least is the sphere of development. The majority of respondents want to solve creative tasks, the minority—unique.
4. The respondent prefers to receive the majority of documents (from 60 to 100%) in electronic form. While the option of reading paper documents is allowed, but electronic sending is preferred.

Based on the results of the survey, it is possible to assume the respondents' expectations of future employment:

1. Employees positively assess the digitalization of business processes, actively use its advantages. Most of the respondents observe the balance of work and personal, do not experience psychological discomfort from digitalization.
2. At the workplace, employees will use messengers and meet face-to-face. The use of phone or mail in the workplace will be reduced. An increase in the use of messengers means that personal and working hours will be less differentiated, colleagues will inform each other if necessary. The desire to meet face-to-face indirectly indicates that the importance of corporate culture will increase in the future.
3. In the labor market, the shortage of development specialists will only increase, the current personnel reserve assesses their development skills as weak, the training process in these areas will play a significant role in the competitiveness of the company, and at the same time it will be especially difficult for respondents. Effective design and implementation of product development programs will be an important challenge for EdTech managers.
4. In their profession, respondents are not ready to face unique tasks, preferring creative ones. It can be assumed that the reluctance to face unique tasks is due to a lack of experience, and as it becomes available, the opinion of respondents may change. The respondents are interested in finding a vacancy in the field of digital professions, i.e. they expect to use as many ICT tools as possible in the workplace. Thus, future cadres support the trends of the labor market.
5. Employees expect interaction with mainly electronic document management, they are ready to read paper documents, but they will send documents in an electronic version. In this regard, in the future, the percentage of electronic document management in companies will be even higher.

4 Discussion

The authors suggest discussing the results again in the future, taking into account the expansion of the contingent of employees and employers. First of all, it seems important to conduct an analysis taking into account the industry and functional specialization of companies. Further, it is interesting to analyze the preferences and level of digital competencies of middle managers, taking into account the functionality of their division.

The study covered employers and employees in St. Petersburg. It seems important to scale up this study and determine whether the preferences of employers and employees differ in different regions of Russia.

5 Conclusion

The conducted research has established the relationship between the preferences of the company's management and the implemented model of digitalization in conditions of high readiness of employees to use digital competencies and their growth.

According to the results obtained, the primary task is to ensure the loyalty of management and understanding of the need to build a unified model of digitalization of business processes of the enterprise. In second place is the task of forming an enterprise team with a high level of digital competencies.

It is desirable to have young specialists able to demonstrate the advantages of using information and communication technologies in all departments of the enterprise.

The problem of formation and development of digital competencies of future employees, including senior and middle managers, is becoming particularly relevant. An important component of the implementation is the development and implementation of programs for advanced training of managers, professional retraining, continuous professional development of managerial personnel.

References

1. Meier, C.: Managing digitalization: challenges and opportunities for business. Management 12, 111–113 (2017). https://doi.org/10.26493/1854-4231.12.111-113
2. Fossen, F.M., Sorgner, A.: Digitalization of work and entry into entrepreneurship. J. Bus. Res. 125, 548–563 (2021). https://doi.org/10.1016/j.jbusres.2019.09.019
3. Kocak, S., Pawlowski, J.: A qualitative study on the categorisation and prioritisation of digital competencies and attitudes for managers and employees. 52–63 (2021). https://doi.org/10.5220/0010674700003064
4. Kozlov, A., Zaychenko, I., Bagaeva, I., Smirnova, A., Glińska-Neweś, A.: The development of industrial enterprise human resources in the process of digitalization: strategic approach.

In: ACM International Conference Proceeding Series (2020). https://doi.org/10.1145/3446434.3446473

5. Schneider, P., Sting, F.J.: Employees' perspectives on digitalization-induced change: exploring frames of Industry 4.0. In: Academy of Management Discoveries, April (2020). https://doi.org/10.5465/amd.2019.0012

6. Nedyalkova, P., Andreeva, A., Yolova, G.: Digitalization and the new legal and economic challenges to employers in implementing internal control. Ikon. Izsled. **30**(5), 158–175 (2021)

7. Haipeter, T.: German trade unions and the digital revolution. Cuad. Relac. Laborales **40**, 301–323 (2022). https://doi.org/10.5209/crla.78214

8. Herron, M.: Social media bullying in the workplace: impacts on motivation, productivity, and workplace culture (2022). https://doi.org/10.4018/978-1-6684-5594-4.ch059

9. Melzer, S.M., Dievald, M.: How individual involvement with digitalized work and digitalization at the workplace level impacts supervisory and coworker bullying in German workplaces. Soc. Sci. **9**(9) (2020). https://doi.org/10.3390/SOCSCI9090156

10. Walter, Y.: The digital transformation in the psychology of workplace spirituality. Digit. Transform. Soc. (2023). https://doi.org/10.1108/DTS-01-2023-0008

11. Malik, N., Tripathi, S.N., Kar, A.K., Gupta, Sh.: Impact of artificial intelligence on employees working in Industry 4.0 led organizations. Int. J. Manpow. **43**(2), 334–354 (2022). https://doi.org/10.1108/IJM-03-2021-0173

12. Eksili, N.: Human resources management: challenges in the digital society. In: Increasing Supply Chain Performance in Digital Society, pp. 262–277 (2022). https://doi.org/10.4018/978-1-7998-9715-6.ch013

13. Juhász, T., Kálmán, B., Tóth, A., Horváth, A.: Digital competence development in a few countries of the European Union. Manag. Mark. **17**(2), 178–192 (2022). https://doi.org/10.2478/mmcks-2022-0010

14. Wilkesmann, M., Wilkesmann, U.: Industry 4.0—organizing routines or innovations? VINE J. Inf. Knowl. Manag. Syst. **48**(2), 238–254 (2018). https://doi.org/10.1108/VJIKMS-04-2017-0019

15. Turja, T., Hakanen, J., Krutova, O., Pertti, K.: Traces of technological well-being: digi-uplifters and digi-downshifters. Nord. J. Work. Life Stud. (2023). https://doi.org/10.18291/njwls.137541

16. Warning, A., Weber, E., Püffel, A.: On the impact of digitalization and artificial intelligence on employers' flexibility requirements in occupations—empirical evidence for Germany. Front. Artif. Intell. **5** (2022). https://doi.org/10.3389/frai.2022.868789

17. Baesu, C., Bejinaru, R.: Knowledge management strategies for leadership in the digital business environment. In: Proceedings of the International Conference on Business Excellence, vol. 14, pp. 646–656 (2020). https://doi.org/10.2478/picbe-2020-0061

18. Amankwah-Amoah, J., Zaheer, K., Woodd, G., Knight, G.: COVID-19 and digitalization: the great acceleration. J. Bus. Res. **136**, 602–611 (2021). https://doi.org/10.1016/j.jbusres.2021.08.011

19. Tartarin, Th., Tichindelean, M., Haaker, T.: Effects of COVID-19 on business models in Romania and the Netherlands, a digitalization perspective. Stud. Bus. Econ. **15**(3) (2020). https://doi.org/10.2478/sbe-2020-0049

20. Hamburg, I.: Impact of COVID-19 on SMEs and the role of digitalization. Adv. Res. 10–17 (2021). https://doi.org/10.9734/air/2021/v22i330300

21. Malterud, K., Siersma, V.D., Guassora, A.D.: Sample size in qualitative interview studies: guided by information power. Qual. Health Res. **26**(13) (2016). https://doi.org/10.1177/1049732315617444

22. Jhantasana, Ch.: Should a rule of thumb be used to calculate PLS-SEM sample size. Asia Soc. Issues **16**, e254658 (2023). https://doi.org/10.48048/asi.2023.254658

Digital Local History and Lore Exploration: Strategies and Tools of Educator's Vocational Competency Development

Andrei A. Bogatyrev and Elena G. Milyugina

Abstract The research topic is predetermined by overwhelming transformations of modern educational activities and vocational competency of school teachers as educative tour guides under present conditions of Digital local history and arts Web 2.0–3.0 representation and exploration development, as well as educative tour initiative booming. *The objective* of the research is to substantiate theory and propose a model of up-to-day productive system of strategies and tools for vocational competency and competitiveness of school teachers as educative tour guides. The *research hypothesis* purports that (a) vocational competency and competitiveness formation of teachers as educative tour guides needs to be directed at multimodal organization, tuning and support of actual educative tour initiatives; (b) the vocational educative tour skills competitiveness of teachers is based on integrate power of vocational hard and soft skills, flexible task-solving strategies, digital literacy and digital educative products making and multimodal communications skills acquisition and development. Within complex and *ecosystem* approach a set of expedient methodological strategies, tools and solutions has been proposed, elaborated and tested. *The method* is based on a cluster approach to competency formation, the multilevel model of vocational educative competency acquisition, a vertical model of increasing the level of competency, associated with an increase in the level of qualifications through the consistent development of digital age competencies of different levels of complexity; and finally—a functional-axiological approach, that reshapes the system of vocational competences based on the principle of tasks priority. As a *result*, the relevant to boosting vocational competency of school teachers as educative tour guides description of educational roadmaps and a systematic approach to different genres of digital educative products has been provided and tested. The set of productive digital educative genres includes: (a) local history and lore information project; (b) interactive digital guide; (c) virtual

A. A. Bogatyrev
University of Science and Technology MISIS, Moscow, Russia

E. G. Milyugina (✉)
Tver State University, Tver, Russia
e-mail: Milyugina.EG@tversu.ru

© The Author(s), under exclusive license to Springer Nature Switzerland AG 2023
Z. Dvořáková and A. Kulachinskaya (eds.), *Digital Transformation: What is the Impact on Workers Today?*, Lecture Notes in Networks and Systems 827,
https://doi.org/10.1007/978-3-031-47694-5_5

local history and lore excursion tour with exploration elements; (d) digital interactive local history and lore quest.

Keywords Digitalization of local history and lore · Educative tour initiative · Ecosystem approach · Digital strategy for the formation of vocational competency and competitiveness · Digital vocational competency of teachers as educative tour guides · Digital educational product

1 Introduction

Digitization of data, digitalization of social communication practices and the global *digital transformation* in the field of local culture, lore and arts representation, research and sharing opens new opportunities for exploration of national cultural heritage across the world and generates new challenges for educational institutions. Meanwhile the whole branch of national tourism is undergoing digital transformation, exploration of cultural heritage in digital age is still concerned with *solving educative tasks*, caring for heritage values and national identity formation issues. *Educative tours* in the specified area are viewed as a goal-seeking interactive process of the travelers' personal familiarization and mastering of the meaningful local geographical, historical and cultural space as decoding new texts and valuable messages. So, these can be viewed within frame of visitor's educative event personal experience.

Such kind of meaningful socialization and inculturation event can be either self-organized with resort to digital advisor [1] or purposefully pedagogically-organized [2], as well as take a field, a virtual [3] or mixt [4] forms of personal educative experience. It generally used to take a standard pedagogically-guided field excursion form in XX century Russia and USSR [5]. Nowadays educative tour initiative has taken over the leading trends in the area. It focuses on travelers' personal encounter and discovery of historical and cultural local sights and events. It lays strong accent on aesthesis of such acts, triggering personal sensitivity, individual's feelings, capacity for empathy, emotional immersion and expanding experience [6]. At the same very time the rise and growth of digital forms of representation of local history and lore tells on the dynamic change of used-to-be educative tour standards as well distribution of the statuses and roles of the participants of educative tourism processes and events. Digitalization process involves online informing and consulting, virtual tours opportunities, all aspects of interacting, management, negotiating and networking of providers of service and customers online [7, 8]. Under Digital Transformation traditional educational institutions are losing their monopoly in the field [9]. The educative tour initiative in the recent years is often taken over by private entrepreneurships, tour operating companies and individuals, venturing to propose private 'authorized excursions' and to perform through the full-fledged educative tour cycle on their own. Internet resources and online world share local attractions online and take on responsibility for organization of trips, excursions and various sorts of educative events in

different corners of the planet Earth. At the same time the development of extracurricular educative tours engagement for schoolchildren is hindered by unpreparedness
of school teachers as educative tour elaborators and tour guides, preconditioning low
level of their competitiveness in such activities [10, 11]. So, the problem of our study
is teachers as tour guides competency formation in pedagogical university.

Here is a set of our research issues—the questions we need to cover against the
background of designing reliable strategies for boosting modern teachers as educative
tour guides competency and competitiveness under the circumstances mentioned.
(1.1.) What way do the challenges of digital local history and lore and initiative
tourism paradigm change the content of the vocational activities of teachers as tour
guides? (1.2.) What vocational competencies should teachers possess and master
in order to keep competitiveness in a multimodal (real and digital) information and
cultural environment? (1.3.) What are the key strategies and tools for the competency
formation of teachers as tour guides?

The study is dedicated to ***the problem*** of eliciting the way and methos for keeping
the vocational competency of modern teachers as educative tour guides at the age of
digitalization of educative trip activities and the spread of initiative tourism paradigm.
The ***theoretical objective*** of the study is scientific description and typology of productive strategies and tools for boosting teachers as educative tour guides vocational
competency. The ***object of the study*** is Digital local history and arts Web 2.0–3.0
representation and communications, causing transformations of educative tours practices. The ***subject of the study*** is specification of productive methodological strategies, tools and solutions for formation and boosting teachers as educative tour guides
vocational competency.

The initial hypothesis of the study is based on the assumption that the vocational
competency of teachers as tour guides in the field of local history, arts, lore and
tourism should develop with respect to the modern context of initiative tourism and
the digitalization of relevant to the field information and services. It is based on the fact
that modern tourist and local history movement in Russia essentially differs from the
organized tourism of the 1930s–1980s in USSR due to *decentralization* of educative
tour organization and management, blurring of tour management hierarchy, and shift
of the previous default focus on an 'average user' towards diversity, introduction of
digitalized and combined/multimodal forms of vocational communication with the
audience of tourists. Taken all this into account, we forward *a hypothesis*:

(a) vocational competency and competitiveness formation of teachers as educative
     tour guides needs to be directed at multimodal organization, tuning and support
     of actual educative tour initiatives;
(b) the vocational educative tour skills competitiveness of teachers is based on
     integrate power of vocational hard and soft skills, flexible task-solving strategies,
     digital literacy and digital educative products making and communications skills
     acquisition and development, as well as mastering a wide range of information
     resources and building operational agility is the adoption of productive digital
     strategies and tools.

   The key research tasks were

- to provide description of the current state, analyze and evaluate the actual challenges in educative tourism;
- to analyze the vocational standard of teachers as tour guide competences in its relation to actual challenges and context;
- to elaborate and test a system of productive methodological strategies, tools and solutions for formation and boosting teachers as educative tour guides vocational competency at digital age.

2 The Methods

(a) What way do the challenges of digital local history and lore and initiative tourism paradigm change the content of the vocational activities of teachers as tour guides? Local history and lore term is introduced here to stand for '*krayevedeniye*', meanwhile krayevedeniye is a term, widely accepted in Russia (and former USSR), and it presently has no distinct equivalent word in English. It stands close to 'regional studies' (Russian '*regionovedeniye*'), targeted at investigation of specifics—a set of meaningful (social, economic, natural, ecological, as well as cultural) features of a definite regions [cf. 12]. Unlike regional studies 'krayevedeniye' addresses a wider span or mix of information and education needs, essentially in ethnographic, educational and educative contexts rather than business, industrial or any other narrow vocational interests of information consumers. Within pedagogical perspective 'krayevedeniye' can be viewed as an important involving activity and resource of formation and development in learners and visitors of such value as "appreciative attitude to nature and ancestral cultural heritage" and its impact on the development of "aesthetic norms and preferences" [13, p. 1, 3]. Another attainable target is the development of pedagogical students essential "value orientations" [14]. So, we admit that including local history, arts and lore research and dissemination as part of 'education content' and vocational skills formation can contribute to vocationalization of university education and students' development of "pedagogic position", viewed as set of axiological goal-guiding values [15, 16]. The implementation of these pedagogical targets nowadays appeals to the XXI century skills development, information, media and technology literacy and skills included [7].

One can indicate a special bundle of information needs, vocational or educational of teachers, researchers, writers, publishers, mass media agents, tour guides, tourist agencies, local cultural life managers and institutions like exhibitions, galleries, libraries, museums, archives, on the one side, and students, tourists, travelers or general audience as consumers of information on the other [cf. 17]. So, we must admit that the very existence of the term 'krayevedeniye' can be justified by actual vast overlapping area for academic investigation of local history, lore and arts, on the one hand, and hospitality industry, tourism, education and public relations, on the other.

There can be specified different stakeholders in development of local history and lore studies. However, the position of our predominant interest is that of educator as cultural missionary. This one can be enacted in the face of a teacher, a museum or art gallery employee, a researcher and a tour guide. 'Krayevedeniye activity' has a strong educative component, concerning the learners' and visitors' information reception and personal response. It can be viewed both as part of general public 'enlightenment' and part of patriotic and civil consciousness education.

The labor functions of an educator tour guide include doing research, relevant information collecting, processing, presenting, and sharing, as well as values dissemination. These presently include digitization of analog type of different kinds of data (previously printed or handwritten documents, pictures, videos, descriptions, interpretations, recommendations, discussions etc.) and sharing the research results in digital format at topic-specific meet-ups, and then on the net. Digitization "makes information more easily accessible, storable, maintained, and shared" [18].

One of ultimate forms of digitization of history, lore, arts and technologies data is presented today in modern digital museums and art galleries. It has a philosophy of its own, addressing the ontological statuses of exhibited digital objects [19]. Meanwhile other researchers pinpoint "user- centered" and "playful" strategy of presentation' of digitized objects [20]. The progressive role of digitization and digital presentation of cultural objects is stressed in optimistic conceptions of "virtual repatriation" and (in other terms)—"… digital repatriation (broadly conceived) as a means to engage local communities, document traditional knowledge, expand the scientific record, and enhance ethnographic knowledge of community practices, languages, and social processes, past and present" [21, p. 14]. Nevertheless, digital local history and lore explorers propagate authenticity of encounters with authentic objects in authentic locations, supported with reliable and appropriate storytelling.

The notion of '*digital local history lore*' (Russian '*tsifrovoye krayevedeniye*') includes and at the same time exceeds the narrow boundaries of already customized term 'digital tourism', meaning "digital support provided to travelers before, during and after the travel activity" [18]. However, we treat 'digital krayevedeniye' as addressing the authenticity of representation issue in quite a different key—the spirit of place should be experienced in offline regime wherever and whenever possible, facilitated by utmost care and support, provided by digital information, made accessible by digital tools and through application of digital technologies—local sanctuaries call for pilgrimages. This approach can be witnessed in part in present-day customers' personal offline navigation of Museums and art galleries in individual tempo, when prompted by narratives, received through headphones, interactive QR-codes, tagged to exhibition items, FAQ-resources, still enjoying offline accessibility of the tour guide's personal presence, support, explanation, guidance and advice. But there is more to it. 'Digital krayevedeniye' opens the precious opportunity for making the tour personal and unique, escaping trite triviality of mass tourism programs and sharing special for a blasé and most demanding visitor. Implementing this approach needs accumulation of data and integration of efforts of different stakeholders of the process, allowing for flexibility of service, variety and attainability of its objectives. Digital records of visitors' feedback also makes a significant part of customer satisfaction monitoring practices within frame of digital history and lore programs development, implementation and improvement. At the same time new business models may occur—like subscription-based model for war history amateur explorers, subscribing to a specified YouTube or some social net channel.

Local history and lore management processes are changing. Digitalization of local history and lore information projects and services opens new opportunities for cooperation of different market players and a vast expansion of different social groups voluntarily involvement, collaboration and contribution to the development of the level of public cultural awareness of local places and objects of cultural heritage and to the design of new high quality educative tours, guidance and products in the area. Application of 'digital platforms reduce the monetary and non-monetary costs of traveling' [22, p. 24]. Digitalization employs *ecosystem approach* in tour offers, providing for involvement of new market players on open online platforms, competition and innovation [23], allowing for elaborating and introducing "less traditional destinations" [22, p. 24]. Digitalization of local history and lore in particular facilitates the processes of elaborating, planning and implementing new touristic routes, based on goal-centered digital resources formation and versatile multipurpose partnership, expedient combination of online and offline forms of communication and interaction. It also boosts tourism as international business, and is a way to implement the key digital transformation slogan—"People are more important than technology" [18].

Digital transformation reshapes the roles of the guides and the excursion group members, on the one side, and poses new challenges, on the other. The demand for exploration is triggered by the customer needs and desire, not limited to a narrow palette of trivial routes in supply any more. The online digital representation of values and remote tour management function as driver of decision-making on choosing the tour, promising a deal of involvement and authentic personal experience. This digital transformation in the area necessitates updating the competences of modern pedagogical university students and graduates, involving both hard and soft skills upgrading.

(b) What vocational competencies should teachers possess and master in order to keep competitiveness in a multimodal (real and digital) information and cultural environment? The answer to present-day local educative tourism needs and challenges is shared between traditional (standardly municipal or academic) official local culture study institutions (universities, libraries, museums, galleries and local press) and the initiative tourism entrepreneurship of 'authorized' tourist programs and services. The resulting quality of educative tourism events depends on the guide's readiness to surf, process and critically analyze different sources, to synthesize a wholesome and exciting narrative and flexibly adapt it to the audience needs and reception capacities.

Meanwhile enthusiast-guides develop their knowledges and skills base on the self-education strategies and according to their own subjective ideas about the set and essence of skills in demand, the comprehensive set of due competences of teachers as educative tour guide is determined by state requirements in the edited official Russian federal document—vocational benchmarking 'professional standard'. The Professional Standard "Teacher (pedagogical activity in the field of preschool, primary general, basic general, secondary general education) (educator, teacher)" (2014, rev.

2016) includes among other such labor actions as "to organize extracurricular cultural and leisure activities, taking into account the place of residence and the historical and cultural identity of the region" [24]. The vocational competence of Russian guides is also described in the Professional Standard "Tour Guide (Guide)" (2014). It includes such labor activities for organizing group and individual excursion activities. However, these activities are presented in the document as essentially homogeneous. The standard mentions mastering modern information technologies and methods of information processing by using modern technical means of communication as part of the competences necessary for the guide. Nevertheless, the application of these in labor activities is not regulated explicitly [25].

The specialized educational programs, implemented in Russian universities (for example, programs of Tver State University 05.03.02 Geography, profile "Recreational geography and tourism"; 43.03.02 Tourism, profile "Technology and organization of tour operator and travel agency services" [26]), despite all their obvious practical orientation, are similarly targeted at preparing graduates to implement mass tourism programs only. The same idea is true concerning the program "44.03.01 Pedagogical Education", which includes in the training of primary school teachers the competence of "the organizer of extracurricular excursion activities" [26]. The observed educational targeting allows us to conclude that the requirements of professional standards for the vocational competence of an educator/teacher as guide and the corresponding models of postgraduates of specialized university programs are insufficiently coordinated with the modern challenges of local history and lore digital representation and initiative tourism. This educational obstacle challenges the competitiveness of university postgraduates in comparison to successful initiative amateur-guides.

The competences of a local history and lore educator are also determined by a special niche, occupied in the excursion service market, which is located between the institutions of large museums and libraries, volunteers' thematic clubs on the one side, and the commercial industry of private initiative entrepreneurship in the field of tourism, on the other. The time resource of a school teacher for the development and preparation of educational routes and educational projects is significantly limited, and the material resources of the exhibition items of school museums can seldom rival the large museums' collections and expositions. At the same time, many genres of commercial and informal authorized stalking programs cannot be applied in school practice—for example, such genres of extreme tourism as rafting and roofing—for security reasons. Moreover, the very promotion of such types of experiencing the region at school can hardly be accepted. Therefore, an important source of attracting and maintaining interest in school routes, along with drawing interdisciplinary connections, are well-developed narratives and hybrid technologies for supplementing artifacts of history and culture with digitized and digital materials proper (video and audio recordings; 3D-modelling, well-structured digital information resources). Digital media contribute to accessibility, replicability, ubiquity, multi-syllabus compatibility of educational materials. Exploiting social media (blogging, advertising, online discussions, landing pages, educational events online coverage) can also contribute to attracting and involving volunteers and experts in local history and lore exploration.

(c) What are the key strategies and tools for the competency formation of teachers as tour guides? The competitiveness of any employee is determined by one's labor potential, associated with good possession of specialized competencies, as well as functional flexibility [27]. The competency of teachers as guides, due to the mixt and interfacing nature of information and organizational support for local history and lore tourism, is predetermined by their ability to integrate the specialized competences of historians, cultural experts, urbanists, museologists, librarians, journalists etc. It involves the educative tour guide's special ability to combine and integrate in agile regime different facts and knowledges into a well-structured, fluent and insightful, coherent and consistent, meaningful and exciting narrative, as well as apply them flexibly in non-standard and changing conditions [28]. Interdisciplinary integration of specialized knowledge and skills allows for effective information support for initiative touring requests in case of insufficient information on the local historical text [29].

In order to form the competency-based competitiveness of tour guides, it is necessary to master modern digital tools and technologies of local history and lore data processing and sharing, as well as educative tour organization and activities support [30]. These can significantly expand the system of specialized digital information resources and communications, contributing to the development of globally accessible digital (virtual and interactive) cultural and educational space of the region.

The educational space of digital local history and lore, which is being purposefully and spontaneously formed today, includes online catalogs and reference books, projects and research practices, digital local history resources of libraries and museums, explorations, performed by of local Internet communities and shared through blogosphere, different projects of research groups and individual enthusiasts, etc. [31] There certainly are some efficient communities and some informative local history and lore portals today. Nevertheless, on the large scale this online sphere is still not systematized and not well-structured as a global information resources chain, which makes a task for the nearest future. However, today's Internet-surfer needs a good guidance and a reliable strategy for mastering and systematizing the digital local history resources and tools offered on the net.

Elaborating such a strategy needs an ***ecosystem approach***. Implementation of ecosystem approach in education is predetermined by plurality of modern educational institutions—official state educational organizations as well as private, a variety of social institutions, Internet communities etc. [32]. Suchlike external factors precondition the productiveness of educational ecosystem and the quality of education [33].

In order to provide for the quality of education and the vocational competency of graduates, the educational ecosystems of universities should be designed as a flexible and adaptive systems of socio-economic relations between universities' own structures and external organizations, targeted at developing scientific, innovative and entrepreneurial projects together with university external partners' organizations [34].

Boosting competitiveness of universities and the graduates depends on the *educational ecosystem*. This system involves the creation of pedagogical conditions for the development of the subjectivity, vocational position, self-determinacy and responsibility of students, based on the organized interaction of the network community of participants, such as students, teachers, administrators, control and monitoring facilities [35], as well as individual researchers and external partnership organizations, interacting through digital infrastructure [36].

Digital educational ecosystem allows for creation of educational and industrial clusters that provide regulation of many flat/horizontal communications, taking forms of research centers, innovation laboratories etc. [37] and enact the participation of university in supporting students' entrepreneurial initiatives [38].

With regard to the task of formation of vocational competency of teachers as educative tour guides, the *ecosystem approach* involves the creation on the basis of this professional interaction of *three* educational and industrial *clusters*: educational (education and research), innovative and entrepreneurial. It engages the network interaction of the university with such partner-organizations as museums, libraries, archives, tourism companies, local history studies associations etc.

The educational cluster, which, in the traditional approach, included the vocational training of teachers at the university and controlling their vocational self-implementation in course of educational practice at school, needs to be organically associated with innovation and entrepreneurial clusters as part of holistic educational ecosystem. Within frame of the educational ecosystem, the vocational activities of students need to receive support from the partner-organizations. Museums, libraries, specialized research centers are called upon to provide support for the initiatives of students in developing local history and tourism digital products in scientific terms, providing them with access to their digital resources and thereby stipulating the development of research activities of students. On the other side, engaging students in digitalization of museums and libraries funds and the development of digital educational products contributes to boosting 'krayevedeniye', the innovation projects and digital production of these partner-organizations. The innovation cluster is inextricably linked with the entrepreneurial one, based on the interaction of the university with tourism organizations and public local history and lore associations. The entrepreneurial cluster provides testing of digital educational products (developed by students) and promotion of them in the field of educative excursion performance and tourism business organization in online and offline regimes.

Thus, within frame of implementation of the educational ecosystem presented, the educational activities of students are supplemented by the formation of a system of local history and lore data and special knowledges and supported by the development of educative tour guide competency in professional cooperation with museums and libraries. The development of pedagogical students' research and digital competencies is stipulated in the process of digitizing analog data and working with archive documents; then digital educative products get introduced in online guides and support and get tested in offline practices.

A set of expedient methodological strategies for development of teachers as educative tour guides vocational competency involves: a cluster approach [39]; the multi-level model of vocational educative competency acquisition, regarding competency formation progress in increase in the of functional flexibility; a vertical model of increasing the level of competency, associated with an increase in the level of qualifications through the consistent development of digital competencies of different levels of complexity [40]; a functional-axiological approach, reshaping the system of vocational competences based on the principle of priority in given situations and contexts [41].

The study is performed within frame of a *complex approach*, allowing for introducing a number of indicators and measurements. The leading role in solving the study tasks belongs to the methods of pedagogical modelling and design, based on introducing elements of ecosystem and immersive approaches. The local history and lore texts of different kind provided material of empirical exploration.

The key expected conclusions and generalizations are aimed at clarifying the ways and tools for the formation of the competency of teachers as tour guides in the context of digitalization of local history and lore objects, routes and phenomena and competing with initiative local history and lore tourism.

Multimodal analysis of objects of local history and lore interest provides opportunity for multifaceted reconstruction of local history sights and culture events, needed for adequate representation in meaningful educative narratives. It also proved useful for development of new educative tour routes, especially not practiced earlier for mass tourist pilgrimage purposes. At the same time, it allowed for tagging the local culture objects according to different special aspects of educative interest and integrating them into larger subject-specific nets of local sights itineraries and either global or local maps of travelers' interest attractions.

The key stages of the study included: input monitoring of the level of formation of *pedagogical students' local history and lore competency*; substantiation of the key priorities of formative work; development of a model of vocational competences and individual *roadmaps* for increasing the level of competency and implementation in a system of new educative routes digital projects; the final monitoring of the level of formation of competency of students as tour guides and the research results analysis.

3   Research Results

The input monitoring results. The testing of the set of developed strategies and tools was conducted in the 2020–2021 academic year in Tver State University (Russia). The study involved 50 graduate students enrolled in 44.03.01 Pedagogical Education (profile "Primary Education"), i.e.—future primary school teachers and 100 junior schoolchildren (grades 3 and 4) of different secondary schools in Tver. The study was conducted by university pedagogical students during the academic semester within frame of the academic discipline "Literary local history". E-learning tools and distance learning technologies and techniques were applied. The educational ecosystem was based on networking with partner educative organizations (such as regional and municipal museums and libraries, the Tver Regional Society of Local Lore, regional tourism associations, etc.). Development of digital information resources on local lore played an important role in the educational ecosystem functionality development.

In the very beginning of experimentation only 20% of the pedagogical students showed possession the core competences of organizing extracurricular activities in a group, micro-group and individual format at a high and 80% of the students at sufficient level. The competence to develop methodological support for extracurricular activities, including in digital format, was detected at sufficient level with 50% of the students, and at low level with another 50% of the students. The knowledge of local history and lore was elicited at a sufficient level with 20% of students and at low level with 80% of students. At the same time the skills for organizing educational excursions were elicited at a sufficient (10%) and low (90%) level. Since the criterion of competency within the framework of this study is the ability to integrate core competencies in excursion work, the initial level of competency was defined either as sufficient (10% of the students) or low (90% of the students).

During their school practice, in order to select topics and determine the targets of the educatory work, the students conducted a remote monitoring of local history knowledges and tourist preferences of the 3d and the 4th grade school-children. As a result, the school-children expressed a desire to get acquainted with the local text of Moscow and St. Petersburg (10%), Tver (10%), smaller ancient cities of Tver region (70%) and historical noble manors and estates (10%). As for selecting educative activities forms, 40% of them preferred cognitive, another 40% preferred mixt cognitive and playing, and 20% voted for mixt educational and research forms of activities.

The design and implementation of educational roadmap for increasing the level of teachers as educative tour guides competency of students. In accordance with the demand, presented by the younger (3d and 4th grade) school learners, the specific topics of core project activities were chosen. The genres of educative activities were specified as (a) local history and lore information project; (b) interactive digital guide; (c) virtual local history and lore excursion tour with exploration elements; (d) digital interactive local history and lore quest. The format accepted was the development of a digital educational product of local history and lore for organizing extracurricular activities in an individual and a micro-group interaction, based on implementation of distance learning technologies. The roadmaps for increasing the level of competency [42] were developed with regard to the key priority targets of formative activity and correlated with the corresponding specified set of tasks and tools mentioned.

The following vocational sub-competences integrated model was proposed and used for testing teachers as educative tour guides competency (Fig. 1).

The project implementation strategy was based on a three-level ecosystem construction, following a cluster approach. The basic (first level) clusters included direct interpersonal interaction of the participants of the project—the pedagogical students, developing their roles as educative tour guides, the schoolchildren as tour

The levels of competences' complexity	The list of sub-competences		
	Core competences (CC)	Interfacing (IC)	Supplementary (SC)
HIGH	high CC	high IC	high SC
MEDIUM	medium CC	medium IC	medium SC
LOW	low CC	low IC	low SC

Fig. 1 The vocational sub-competences integrated model for testing teachers as educative tour guides competency formation

initiators and the school teachers. Based on expressed tourist preferences, the micro-groups of schoolchildren and students chose a location in Tver region (such as Tver, Torzhok, Staritsa, Vyshny Volochek, Kimry, Kalyazin etc.), formulated the key tasks of the project and distributed tasks for schoolchildren and their teachers' performance.

In order to clarify and systematize the collected digital data, second-level clusters were organized. The latter made it possible to implement intra-university interaction of pedagogical students' project participants with academic professionals and fellow-students of different other educational profiles (philologists, folklorists, historians, geographers, culturologists) and thereby to receive a professional support for specific modules of the project. Further on, in order to transfer the project from a virtual format to an offline and combined one a set of the third level clusters was introduced. The latter one was focused on the interaction of project participants with external institutions-partners, such as libraries, museums, local history communities etc.

(a) Local history and lore information project. The local history and lore digital information project genre addressed the objectives of forming the core competencies of students as basic ones (the low CC level of complexity, Fig. 1). Its implementation contributed to the formation of a set of systematic local history and lore knowledges of students. A cluster approach was used for the task implementation. A micro-group of young learners in charge of the pedagogical institute students made their own choice of a local history and lore text, anchored to a specified geographical object, located in Tver region (such cities as Tver, Torzhok, river Volga, lake Seliger, etc.). Then the coverage focus of the report was specified—for example, "Volga through the eyes of a travel-reporter", "Tver, reflected in a local famous writer correspondence heritage" etc.). The digital resources involved in the project were the digital libraries "Tverskoy Digest" [43], "Tverskoy Krai" [44], etc.

(b) Interactive digital guide. The development of this educative product was focused on mastering the core competencies with a certain increase in the level of complexity according to the vertical principle "low CC—medium CC—high CC level" (Fig. 1). The elaboration was based on horizontal connections, including popular science, folklore and fiction texts, relevant to spatial, temporal and topical criteria. Such multi-modal approach allowed to integrate the core pedagogical competences of students with interfacing folklore and philological ones. In the process of preparing an interactive digital guide a micro-group of schoolchildren together with pedagogical students developed some initiative and authorized local history and lore educative tour routes (such as "Tver to Staritsa", "Navigating Volga from Rzhev to Tver", "Historical estates of lake Seliger neighborhood" etc.). The values of the routes had also been discussed and established in teamwork. The footages of the project were found in the digital resources of the collection of the Tver Gorky Academic Library [45].

(c) Virtual local history and lore excursion tour with exploration elements. The development of this digital educational product was aimed at the consistent development of core (CC), interfacing (IC) and supplementary (SC) competencies by using a vertical method of increasing the level of teacher as guide competency (Fig. 1).

In developing a virtual tour, the younger learners, having previously formed their basic local history and lore knowledges and skills as a result of the implementation of the first two projects, chose a certain local history and lore text and designed the corresponding educative route for implementing an initiative educative tour program.

In addition to surfing literary works resources for useful information selection and systematizing, students procured relevant visual sources of information (photographs, paintings and other sorts of graphic materials), as well as expanded the range of folklore texts processing, including urban legends; collected oral stories of local residents. The integrated sources analyses contributed to synthesizing the core, interfacing and supplementary competences in the target competency formation and development. Special *gamization* of cultural immersion and new experience (role-played) task-solving formation *techniques* were introduced. For example, it was solving some tzarina's court communication charades through decoding the secret language of fans in Tver Imperial Palace. The distinct vocational competences of literature and art criticism, culturological study and urbanist exploration were involved in the productive target competency merge for such initiative tours projects as "Tver in Peter the Great epoch"; "Torzhok perceptions through the eyes of travelers of the XIX century"; "Kalyazin—the Russian Atlantis, witnessed in painting and photography" etc.). Boosting the effectiveness of the projects was achieved through expansion of data collection and processing, based on engaging the digital resources of the collections of Tver museums [46–48] and local history and lore communities [44, 49].

(d) Digital interactive local history and lore quest. Digital interactive quest was used as a tool for integrating previously formed core (CC), interfacing (IC) and supplementary (SC) competences with targeted increase in the level of complexity to the advanced level, specified in the model matrix (Fig. 1) and flexible systematization of discrete competences with regard to the principle of priority value. In developing an interactive digital quest all the previously mastered digital resources were systematically engaged, and the various challenging points of Tver local history (puzzles, enigmas; losses, information gaps, conflicting versions) were updated and (whenever possible) resolved. It made possible to include into the targeted digital interactive local history and lore quest some exciting heuristic elements of local history, exploratory games and gamification elements. The quest topics implemented included "Tver Sloboda", "Tver merchant household", "Tver nobility estate", "Tver traditional costume", "Tver cuisine" and other. The genre and format of the quest functioned as an optimal response to the initiative requests of young learners, concerning exploration of minor rich in history towns and noble estates of Tver region.

4 Discussion

The final monitoring elicited significant dynamics in the level of formation of pedagogical students' teacher-as-educative-tour-guide competency. This result was facilitated by the integration of the competences of a teacher-as-guide (the final level of formation was identified as high 80% and sufficient 20% of pedagogical students) and as a developer of a digital educative product of a local history and lore exploration and educative tour (the final level of formation was assessed as high with 70% and as sufficient with 30% of pedagogical students). The progress in functional flexibility development was also facilitated and observed in the project implementation process.

In the course of the local educative tourism information resource and organizational support for an *educative tour digital product development* the pedagogical students mastered the method of multimodal analysis of urban, rural and estate local history, lore and arts texts. They also learned to specify nontrivial educative tour routes and their geopolitical, historical, aesthetic and socio-cultural values and elaborate strategies of their methodological implementation [50–52].

Digital transformation of local history and lore ('tsifrovoye krayevedeniye') is new social reality, overthrowing the used to be stereotypes and calling to use new opportunities. The competitiveness of national educative values at the local history and lore market depends on the successful formation of relevant competency of university graduates and school teachers. In the course of pedagogical experiment pedagogical students' *digital skills* were enacted and tested in digitizing relevant information from rare and archive analog sources, in procuring digital elements in designing guides and visuals for new educative routes. Pedagogical students' *digital practices skills* were enacted in elaborating meaningful narratives and methodological support for tour guides and visitors. The completion of competency formation involved synthesis of the whole range of twelve XXI century skills [7], including modern learning skills, information literacy skills and life skills.

All in all, the hypothesis of the study proved productive. The multimodal organization, tuning and support of actual educative tour initiatives by pedagogical students proved productive within frame of ecosystem approach and learning strategies, implementing vocational competency formation and development road maps. It was based on integrate power of vocational hard and soft skills, flexible task-solving strategies, digital literacy and digital educative product-making and communications skills acquisition and development, as well as covering and processing a wide range of information resources and mastering digital strategies and tools in good balance with offline digitally driven involving and competitive educative practices.

5 Conclusions

Based on the theoretical and empirical research done and presented in the chapter, a set of expedient methodological strategies, tools and solutions has been proposed, elaborated and tested. The targeted competency formation is nested in educational ecosystem. It includes a cluster approach to competency formation; the multilevel model of vocational educative competency acquisition, regarding competency formation progress in increase in the of functional flexibility; a vertical model of increasing the level of competency, associated with an increase in the level of qualifications through the consistent development of digital competencies of different levels of complexity; a functional-axiological approach that reshapes the system of vocational competences based on the principle of priority value in certain situations and contexts. The relevant to boosting vocational competency of school teachers as educative tour guides description of roadmaps and different genres of digital educative products has been specified. The set of tested genres includes (a) local history and lore information project; (b) interactive digital guide; (c) virtual local history and lore excursion tour with exploration elements; (d) digital interactive local history and lore quest.

The overall generalization is that digital media is not hostile to humane experience tradition, when both treated with care. The applied prospects of further local lore research concern further digitalization, processing, systematization and representation of Tver region local history and lore data as tool of educative tourism management and integration of significant information in global historical and cultural heritage semantic net. The prospects of further pedagogical research concern an in-depth study of the interface of digital and life competences formation and implementation.

References

1. Erbil, E., Wörndl, W.: Personalization of multi-day round trip itineraries according to travelers' preferences. In: Stienmetz, J.L., Ferrer-Rosell, B., Massimo, D. (eds.) Information and Communication Technologies in Tourism: Proceedings of the ENTER 2022 eTourism Conference, January 11–14, 2022, pp. 187–199. Springer, Switzerland (2022)
2. Naumov, N., Green, D.: Mass tourism. In: Jafari, J., Xiao, H. (eds.) Encyclopedia of Tourism, pp. 594–595. Springer, Switzerland (2016)
3. Verkerk, V.-A.: Virtual reality: a simple substitute or new niche? In: Stienmetz, J.L., Ferrer-Rosell, B., Massimo, D. (eds.) Information and Communication Technologies in Tourism: Proceedings of the ENTER 2022 eTourism Conference, January 11–14, 2022, pp. 28–40. Springer, Switzerland (2022)
4. Buhalis, D., Karatay, N.: Mixed reality (MR) for generation z in cultural heritage tourism towards metaverse. In: Stienmetz, J.L., Ferrer-Rosell, B., Massimo, D. (eds.) Information and Communication Technologies in Tourism: Proceedings of the ENTER 2022 eTourism Conference, January 11–14, 2022, pp. 16–27. Springer, Switzerland (2022)
5. Arcybashev, D.V., Arcybasheva, T.N.: Detsko-yunosheskaya turistsko-ekskursionnaya deyatel'nost' v Sovetskom Soyuze: gosudarstvennaya politika i formy organizacii. Uchenye zapiski: Elektronnyj nauchnyj zhurnal Kurskogo gosudarstvennogo universiteta 3(51), 1, 1–10 (2019)

6. Al-Ababneh, M., Masadeh, M.: Creative cultural tourism as a new model for cultural tourism. J. Tourism Manag. Res. **6**(2), 109–118 (2019)
7. Stauffer, B.: What Are 21st Century Skills? https://www.aeseducation.com/blog/what-are-21st-century-skills. Accessed 11 July 2023
8. Xiang, Z., Fuchs, M., Gretzel, U., Höpken, W. (eds.): Handbook of e-Tourism. Springer, Switzerland (2022)
9. Egorova, E.N.: Problemy razvitiya sistemy podgotovki kadrov dlya industrii turizma i gostepriimstva. In: Kryukova, E.M. (edit.) Turizm i gostepriimstvo skvoz' prizmu innovacij, pp. 29–32. Universitetskaya kniga, Moskva (2018)
10. Kozina, E.F.: Mesto ekskursovedeniya v sisteme podgotovki uchitelej nachal'nyh klassov. Sovremennye problemy nauki i obrazovaniya **6**, 1–9 (2020)
11. Merzlyakova, L.A.: Problemy podgotovki kadrov v sferah detsko-yunosheskogo turizma i detskogo otdyha. In: Vervinskaya, N.V., Procenko, L.M. (eds.) Turistsko-kraevedcheskaya deyatel'nost' v regionah Rossii: opyt—problem—perspektivy, pp. 26–30. RIC BIFK, Ufa (2021)
12. Gladkij, Ju.N., Chistobaev, A.I.: Regionovedenie: uchebnik dlja vuzov. 2-e izd., pererab. i dop. Jurajt, Moskva (2023)
13. Artaeva, G.V., Appaeva, Ya.B., Kyunkrikova, I.V., Mutyrova, A.S., Shovgurova, I.V., Boltyrova, T.M.: Formation of artistic and aesthetic culture of students in rural social environment. In: SHS Web of Conferences 97, 01025, pp. 1–8 (2021)
14. Zulharnaeva, A.V.: Pedagogicheskie usloviya realizacii vospitatel'nogo potenciala kraevedeniya s cel'yu razvitiya grazhdanskoj identichnosti lichnosti. Nizhegorodskoe obrazovanie **3**, 74–81 (2022)
15. Bogatyrev, A.A., Vanchakova, N.P., Chernyavskaya, A.P., Krasilnikova, N.V., Vatskel, E.A., Babina, A.A.: Development of a vocational and pedagogic position reflection of a doctor as part of continuing medical education program. In: International Conference "Technological Educational Vision" (TEDUVIS 2020), vol. 97, pp. 1–10 (2021)
16. Vanchakova, N., Bogatyrev, A., Denishenko, V., Krasilnikova, N., Shaporov, A., Vatskel, E.: Pedagogic position of resident physicians as a factor contributing to forming a healthy-oriented lifestyle in patients. In: BIO Web of Conferences 29(4), 01023, pp. 1–8 (2021)
17. Mamontov, A.V., Scherba, N.N.: Krayevedcheskaya bibliographia. Kniga, Moskva (1989)
18. Malak, H.A.: Digitization vs Digitalization: What's The Difference? June 18, 2023, https://the ecmconsultant.com/digitization-vs-digitalization/. Accessed 11 July 2023
19. Geismar, H.: Museum Object Lessons for the Digital Age. University College, London (2018)
20. McCarty, C.: Introduction. In: Making Design, pp. 26–31. Cooper Hewitt, Smithsonian Design Museum, New York (2014)
21. Bell, J.A., Kimberly, C., Turin, M.: Introduction: after the return. Museum Anthropology Review **7**(1–2), 1–21 (2013)
22. López-Córdova, J.E.: Digital Platforms and the Demand for International Tourism Services (February 13, 2020). World Bank Policy Research Working Paper No. 9147. The World Bank, Washington (2020)
23. Bogolyubov, V.S., Bogolyubova, S.A.: The potential of using digital platforms in tourism. Vestnik Altajskoj Akademii Ekonomiki i Prava **4**, 156–162 (2023)
24. Professional'nyj standart 01.001 Pedagog (pedagogicheskaya deyatel'nost' v sfere doshkol'nogo, nachal'nogo obshchego, osnovnogo obshchego, srednego obshchego obrazovaniya) (vospitatel', uchitel'), https://classinform.ru/profstandarty/01.001-pedagog-vospitatel-uchitel.html. Accessed 11 July 2023
25. Professional'nyj standart 04.005 Ekskursovod (gid), https://classinform.ru/profstandarty/04.005-ekskursovod-gid.html. Accessed 11 July 2023
26. Informacija o realizuemyh obrazovatel'nyh programmah TvGU, https://tversu.ru/sveden/education/eduaccred/. Accessed 11 july 2023
27. Nikolaev, N.A., Poleshchuk, M.N.: Metodicheskij podhod k opredeleniyu i ocenke konkurentosposobnosti personala. Vestnik Yuzhno-Ural'skogo gosudarstvennogo universiteta: Ekonomika i menedzhment **15**(1), 93–109 (2021)

28. Lozickij, V.L.: Evolyuciya kompetentnostnoj sostavlyayushchej v professional'noj podgo-tovke specialistov v sfere turizma i gostepriimstva v aspekte processov cifrovizacii. Turizm i gostepriimstvo **2**, 34–40 (2019)
29. Afanas'eva, A.V., Logvina, E.V., Hristov, T.T.: Metodologicheskie osnovy nauchno-populyarnogo turizma. Servis v Rossii i za rubezhom **17**, 2(104), 5–25 (2023)
30. Gomilevskaya, G.A., Majorova, E.A.: Innovacionnye formy ekskursionnogo obsluzhivaniya v kontekste pedagogicheskoj funkcii kraevedeniya. Azimut nauchnyh issledovanij: pedagogika i psihologiya **10**, 3(36), 76–79 (2021)
31. Shevyreva, E.: "Digital-kraevedenie 2020": kak izuchat' i sohranyat' istoriyu goroda s pomoshch'yu cifrovyh instrumentov, 21.12.2020, https://news.itmo.ru/ru/education/cooper ation/news/9993/. Accessed 11 July 2023
32. Rozin, V.M.: Ekosistemnyj podhod v obrazovanii—optika, analitika, proektirovanie, transfor-maciya v napravlenii budushchego. In: Shalashova, M.M., Shevelyova, N.N. (eds.) Nepre-ryvnoe obrazovanie v kontekste Budushchego: sbornik nauchnyh statej po materialam IV Mezhdunarodnoj nauchno-prakticheskoj konferencii, pp. 8–18. A-Prior, Moskva (2021)
33. Bandyopadhyay, S., Bardhan, A., Dey, P., Bhattacharyya, S.: Education Ecosystem. In: Bridging the Education Divide Using Social Technologies, pp. 43–75. Springer, Singapore (2021)
34. Izotova, A.G., Gavrilyuk, E.S.: Ekosistemnyj podhod kak novyj trend razvitiya vysshego obrazovaniya. Voprosy innovacionnoj ekonomiki **12**(2), 1211–1226 (2022)
35. Shaporov, A.M.: Pedagogicheskie uslovija akademicheskoj uspeshnosti obuchajushhihsja, 5.8.1. Obshhaja pedagogika, istorija pedagogiki i obrazovanija (pedagogicheskie nauki), Avtoref. dis. ... kand. ped. nauk. Jaroslavl' (2022)
36. Neborskij, E.V.: Cifrovaya ekosistema kak sredstvo cifrovoj transformacii universiteta. Mir nauki: Pedagogika i psihologiya **4**(9), 1–11 (2021)
37. Isaeva, A.E.: Innovacionnaya cifrovaya obrazovatel'naya ekosistema kak baza perekhoda k industrii 4.0. Gosudarstvennoe upravlenie: Elektronnyj vestnik **96**, 177–192 (2023)
38. Moraes, G., Fischer, B., Campos, M.L., Schaeffer, P.: University ecosystems and the commit-ment of faculty members to support entrepreneurial activity. BAR—Brazilian Administr. Rev. **2**, 1–26 (2020)
39. Davydenko, T., Zhiljakov, E.: O klasternom podhode k formirovaniju professional'nyh kompetencij. Vysshee obrazovanie v Rossii **7**, 69–76 (2008)
40. Merkulova, L.P.: Kompetentnostnaja model' professional'no-mobil'nogo specialista tehnich-eskogo profilja. Vestnik Samarskogo gosudarstvennogo ajerokosmicheskogo universiteta im. akademika S.P. Koroljova (nacional'nogo issledovatel'skogo universiteta) 1, 294–303 (2006)
41. Gubarenko, I.V.: Rol' aksiologicheskogo podhoda v formirovanii professional'noj kompe-tentnosti vypusknika obrazovatel'noj organizacii vysshego obrazovanija. Nauka, Iskusstvo, Kul'tura **1**(5), 186–190 (2015)
42. Miliaeva, L.G., Bavykina, E.N.: Upravlenie konkurentosposobnostiu personala organi-zatcii v usloviiakh realizatcii kompetentnostnogo podkhoda. Vestnik Novosibirskogo gosu-darstvennogo universiteta jekonomiki i upravlenija **4**, 27–34 (2014)
43. Tver Digest, https://otveri.info/. Accessed 11 July 2023
44. Tver Region/Tver regional history and arts community Tverskoi krai/Tverskoe oblastnoe kraevedcheskoe obshchestvo, https://tverkray.ru/. Accessed 11 July 2023
45. Tverskaia oblastnaia universalnaia nauchnaia biblioteka im. A.M. Gorkogo, http://opac.tverlib. ru/opacg/. Accessed 11 July 2023
46. Muzej tverskogo byta, Tver', https://tvermuzeum.ru/affiliates/mtb. Accessed 11 July 2023
47. Tverskaja oblastnaja kartinnaja galereja, http://gallery.tver.ru/. Accessed 11 July 2023
48. Tverskoi gosudarstvennyi ob'edinennyi muzei, https://tvermuzeum.ru. Accessed 11 July 2023
49. Associacija chastnyh i narodnyh muzeev Rossii, https://www.chastnyemuzei.rf/o-proekte/geo grafiya/tverskaya-oblast/. Accessed 11 July 2023
50. Milyugina, E.G., Kuzminova, I.A., Rudenko, A.A.: Nravstvenno-patrioticheskoe vospi-tanie mladshih shkol'nikov na kraevedcheskom materiale: kvest-putevoditel' "Kimry: gorod bashenok i teremkov". In: Lel'chickij, I.D. (edit.) Tradicii i novacii v professional'noj podgotovke i dejatel'nosti pedagoga, vol. 19, pp. 282–288. TvGU, Tver' (2021)

51. X regional'naya nauchno-prakticheskaya konferenciya "Kul'tura Tverskogo kraya i so-vremennoe obshchestvo": programma, 16.04.2021, https://pedfak.tversu.ru/pages/1647. Accessed 11 July 2023
52. X regional'nyj konkurs issledovatel'skih i tvorcheskih rabot studentov "100 mest Tverskogo kraya, kotorye stoit uvidet'": programma, 16.04.2021, https://pedfak.tversu.ru/pages/1647. Accessed 11 July 2023

Peculiarities of the Application of Cognitive Learning Systems in the Study of Electrical Engineering Disciplines

N. N. Tsybov⬚ and Zh. T. Galbaev⬚

Abstract The relevance of the considered problems is due to the low didactic efficiency of the application of automated information educational resources, as well as the increased interest of electrical universities to the capabilities of information learning systems when modeling processes in electrical circuits. The aim of this chapter is to improve the quality of the learning process when applying information learning systems, taking into account the psycho-factors of the participants in the educational process. The methodological basis for analyzing problems in the design and application of informative teaching systems is an integrated application of complementary approaches, which considers the information educational environment as a system containing a number of interconnected subsystems. By the example of the developed informational learning system and the virtual electronic training devices included in it, the chapter presents a variant of implementing electronic educational resources taking into account the psycho-factors of the educational process participants. The psycho-factors influencing the efficiency of the learning process are identified by the learning system as a result of diagnosing the personality traits of the students. The identified psycho-facts are used by the information learning system when adjusting the presentation of the complexity of educational material and at the same time, the psycho-factors are additional elements of system analysis—cognitive elements of psycho-factors. As an example of application of the simulator virtual model of a precision power supply system as a part of the information learning system, the paper presents a calculation-experimental method for designing a precision DC voltage regulator.

Keywords Precision voltage stabilizer · Cognitive learning systems · Systems analysis · Psychofactors · Didactic efficiency

N. N. Tsybov (✉) · Zh. T. Galbaev
Kyrgyz State Technical University n. a. I. Razzakov, Bishkek, Kyrgyz Republic
e-mail: nikolay_research@mail.ru

Z. Dvořáková and A. Kulachinskaya (eds.), *Digital Transformation: What is the Impact on Workers Today?*, Lecture Notes in Networks and Systems 827,
https://doi.org/10.1007/978-3-031-47694-5_6

1 Introduction

1.1 Topicality of the Problem

All technical fields of industry use specialized electronic devices, the design of which requires design engineers of the appropriate level of professional training. The task of didactically effective application of cognitive information training systems remains unresolved in the process of implementing the task of increasing the efficiency of education and improving the quality of engineering design personnel training. The main problem of existing information educational resources is their low didactic efficiency. The automation of the learning process can be effective only if the automation will pick up real effective methods of training and education. The effective use of information and educational system containing didactic components requires an educational environment, which includes the main value orientations, as well as a teacher, who is a carrier of such cultural values. In the present research on the example of the developed informational learning system methods of increasing didactic efficiency of informational learning systems application, taking into account psycho-factors influencing the processes of training and education are considered.

1.2 Literature Review

Analysis of the scientific literature over the past two decades has shown that the issue of automating the diagnosis of personality traits of participants in the educational process is poorly developed and little researched.

The most discussed topic in the analysis of the educational process is the problem of education quality crisis. S. Yu. Polyankina correlates the causes of the crisis of education with distortions in the essence of educational evaluation criteria [1]. The same opinion and V. E. Chernikova, noting incorrect understanding of the essence of spiritual and moral values [2].

The basis of the success of any method of teaching depends on the presence of cognitive motivation of students. This issue was studied by V. V. Andreev, who proposed methods of increasing motivation [3]. The interrelation of basic needs with akakdemic performance and with motivation to learn is considered in the works of Campbell and Erten [4, 5].

One of the tasks of implementing the goals of improving the quality of engineering training is to increase teachers' innovative readiness for self-development. The goals and objectives of teachers' self-development are of course relevant, but it is worth noting that under self-development most teachers of technical universities understand professional development in the particular discipline taught. At the same time, the didactic part of the educational process is neglected [6, 7].

The analysis of the psycho-physiological state of the trainees for the purpose of adaptation with electronic means of learning was carried out by V. M. Glushan's works. In her works the model taking into account the psycho-physiological characteristics of the learner is proposed [8, 9]. But it should be noted that the analysis of psycho-emotional state does not fully reflect the personal characteristics of the learner. Only the analysis of ego-motifs—goal motifs and situational motifs, conditioned by superego, can give a real picture of the learner's state. The issues of the learner's adaptation to educational electronic resources are reflected in M. V. Lagunova's work. M. V. Lagunova suggests that the problem of adaptation to automated electronic resources should be solved by interpersonal communication between students and instructors [10].

Analysis of the possibility of complete replacement of the teacher with electronic learning tools is conducted by A. N. Shishkov, who proved the advantage of joint interaction between the teacher and electronic educational tools [11].

Relevant research is the work of D. A. Pechnikov, who recommends taking into account the anthropotechnical component of the problem of didactic interaction of the student with the learning system when constructing an information system [12].

One of the problems in designing information learning systems is the lack of a unified mathematical approach in formalizing the educational components. One of the successful variants of the effective approach to formalization of educational components in describing the model of expert system of educational process quality is the study of Gordienko [13]. The synergetic approach to formalizing the models of educational process is well-proven in the work of Kramov [14].

A review of the scientific literature has shown that the main emphasis of research in the field of psych didactics is focused on the problem of the crisis of the quality of education in general. At the same time, the problem of unsatisfactory didactic efficiency of applied automated learning tools for technical universities remains without due attention. The algorithms of the existing learning systems do not take into account the person-centered approach in education, which requires automating the process of diagnosing the personal qualities of the participants in the educational process. As for the issue of enhancing teachers' innovative readiness for self-development, the overwhelming majority of teachers at technical universities when upgrading their qualifications give preference to improving their knowledge in the field of the subjects taught, leaving the problem of upgrading skills in the field of mastering new didactic methods of knowledge transfer unattended. Therefore, in order to really improve the efficiency of the educational process, software and hardware educational resources, algorithms of which contain didactic principles of teaching, educational environment and a teacher, who is a bearer of new didactic knowledge, are needed.

2 Materials and Methods

Due to the nonlinearity of the learning process, and the uncertainties of pedagogical situations, the methodological basis for analyzing problems in the design of information learning systems is an integrated application of complementary approaches, considering the information educational environment as a system containing a number of interrelated subsystems.

Decomposition, aggregation, and analysis methods were used to determine the composition of the components of the educational process, their causal relationship, structural relationships, and the attributes and qualities of the identified components. Cognitive analysis was used to identify weakly structured components.

In clarifying the composition and patterns of functioning of the elements, as well as in determining the quantitative and qualitative characteristics of homogeneous components of the educational process, structural analysis and the method of comparison were used. When studying the components of technical means, modeling methods were used.

For the purposes of research in the field of psychodidactics the following types of system analysis were used—retrospective, statistical and situational analysis.

3 Results

According to the main functional purpose of automated information educational tools perform the role of analysis and management of the educational process of training and education. Learning systems are hardware and software technical means, algorithms which, in addition to information about the subject area and the means of monitoring training, contain information about the basic pedagogical concepts of training and education.

In the present research, when formalizing the educational components of the learning system, a synergetic approach is used, in which the self-organizing educational environment using e-learning is represented in the form of an ergatic model "learner—information learning system".

The synergetic approach to modeling the educational process is characterized by a certain compression in describing the models, as it describes not the entire system, but the order parameters of the learning process. In our case, the order parameter is memory [14].

In synergetic approach to modeling of educational process the ergatic model "learner—information learning system" is described by differential equations with lag:

$$\frac{dX(t)}{dt} + kX(t - \tau_T) = d\phi \tag{1}$$

where: $X(t)$—vector characterizing the system at time t; d—information flow; k—parameter characterizing knowledge perception; t—time; τ_T—dynamic lag vector.

Monitoring of the quality of the educational process in the proposed informational learning system is organized according to the method proposed by Gordienko [13].

Features of information learning systems.

Let's consider the peculiarities of training systems on the example of functioning of the information learning system developed by us.

The learning system presented in this research has the following distinctive features:

1. Due to the unpredictability of pedagogical situations learning system allows the teacher to quickly make adjustments to the components of the subject area and adjust the type of teaching material by the results of the personal characteristics of the student.
2. In order to activate cognitive motivation in learning knowledge, the database of the learning system contains training tasks, as close as possible to the implementation of tasks in the profile of the future engineering activity of students.
3. In order to implement a person-centered approach in education, the system takes into account psycho-factors that influence the educational process and forms the type, volume, and trajectory of the training material. For this purpose, the training system includes a psychodiagnostic module of personality traits of the participants in the educational process. In addition to the main function of psychodiagnosis, the diagnostic results are used by the information learning system in the formation of new additional cognitive elements of psychofactors, which makes it possible to obtain a system analysis with new qualities.
4. The means of visual information display of the training system has the ability to adjust the color design of information blocks on the monitor according to the psychotype of the learner.
5. The information training system includes electronic training devices that take into account students' personal characteristics. In the training system training devices, taking into account psycho-factors also play the role of learning level sensors and indicators of the control object state, which is a student [15].
6. The information training system uses new author's methods of psychodiagnostics and students' consciousness attunement to the perception of training material [16].

The activity of engineers designing electronic devices involves a number of stages of calculation and experimental works. The existing methods of calculating the parameters of electrical circuits using semiconductor components have a small accuracy of $10–15\%$ in engineering calculations. Such a low accuracy is due to the wide variation of parameters of passport data on electronic radio components.

To refine the engineering calculations in the present research, a comprehensive engineering method of electrical circuit calculation with the application of cascade modeling of processes in the designed circuits is proposed in the design. With such a comprehensive approach, it is possible to increase the calculation accuracy of the designed circuits by at least an order of magnitude.

Students of electrical engineering are particularly interested in the design of electronic devices which they will apply in their future engineering activities. Therefore, cognitive training systems should include virtual simulator models, which allow the design of components of electronic assemblies of real devices.

Any electronic device is not without power supply system, so the development of stabilized power supply systems is an urgent task. For the purposes of maintenance of series production, designing stabilized power supply systems is not difficult. But for scientific research in the study of processes in electric circuits in some cases precision power supply systems with output voltage stability in thousandths of a percent are required. Therefore, virtual simulation models of precision stabilized power supply are of particular interest to students of electrical engineering.

Let us consider an example of calculation and simulation of a precision power supply system using a simulator device of an information training system.

The training device, created on the basis of a precision voltage stabilizer, in addition to the main node, which ensures the functioning of the device, contains redundant nodes of varying complexity in the design. The training device of the stabilized power supply system contains 6 main nodes, each of which has a variable part in the form of similar functionality modules.

In the process of interaction of the student with the training system according to the results of academic testing and the results of psychodiagnosis of the student's personality, the information learning system offers to design a variant of the node with the complexity corresponding to the student's training.

If the student is asked to design a precision power system with output voltage stabilization accuracy of hundredths of a percent, the student will be asked to design a variant of a voltage reference node built on two counter-coupled current stabilizers that feed the main reference DC current regulator. And in the case of designing a stabilizer to power digital microcircuits, the student will be asked to use a single stabilizer and a resistor as the voltage reference.

In the case of designing the output and pre-terminal stages of a voltage regulator for a load current of 50 A the student will be asked to design a composite output stage, and in the case of designing a stabilizer with a load current of 0.5 A there is no need for both the composite output stage and the pre-terminal stage.

Let's consider an example of an experimental and computational method of designing a precision stabilized power supply system, with an output voltage that is virtually independent of the load current. The student is asked to design a DC voltage regulator with an acceptable deviation of the output voltage when the load current changes *0.001%*.

When designing a precision stabilizer with an output voltage of 12 V 50 A and with a stability of at least *0.001%*, the question of system stability to excitation must be solved, since such accuracy requires a high stabilization factor [17].

When designing, the student should make a general functional diagram of the designed device and perform a step-by-step calculation of all the nodes proposed for design with cascade modeling (see Fig. 1).

To calculate the circuits of the stabilizer it is advisable to start with the calculation of the parameters of the output stage.

To select the transistors of the output stage it is necessary to calculate the output power of the stabilizer load:

$$P_{LOAD.} = U_{OUT.} \times I_{LOAD.} = 12 \times 50 = 600 \, \text{W} \tag{2}$$

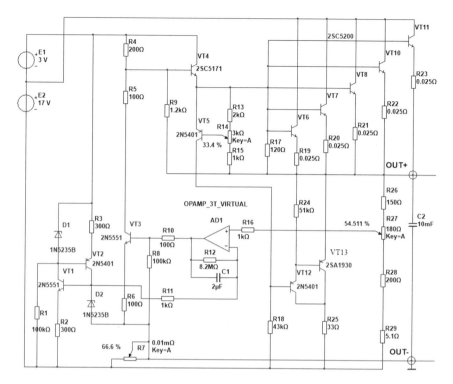

Fig. 1 Schematic diagram of a precision voltage regulator

where: $U_{OUT.}$—is the rated output; $I_{LOAD.}$—is the maximum load current.

At power in the load of 600 W, *60–70%* of the power in the load, that is 350–400 W will be dissipated on the output transistors. Therefore in order to choose the optimum area of heat sinks for the output stage five transistors with a maximum power dissipation of 150 W (2SC5200) must be used.

Let's determine the values of nominal voltages of the main supply voltage E_2 and the value of volt-switching voltage E_1 (see Fig. 1).

Let's set the voltage E_1 for the supply of the pre-terminal stage to 3 V.

The minimum value of the main supply voltage E_2 in this case can be determined from the expression:

$$U_{E2MIN.} = U_{OUT.RV.} + \Delta U_{R19} + \Delta U_{CEVT6} = 12 + 0.25 + 1.35 = 13.6\,\text{V} \quad (3)$$

where: $U_{OUT.RV.}$—rated output voltage of the regulator, equal to 12 V; ΔU_{R19}– voltage drop across *R19* at maximum load current; ΔU_{CEVT6}– minimum voltage between collector and emitter when the transistor has not entered full saturation.

According to the characteristics of the SC5200 transistor, the saturation voltage at a collector current of 10 A is 0.4 V. To keep the transistor still active and not in saturation, we will take 1.35V as the minimum voltage between collector and emitter.

For a given input voltage deviation of *20%*, the input voltage variation range will be from 13.6V to 20.4V at a nominal input voltage of 17 V.

Considering the variation of commercially available components in parallel connection, it is necessary to apply a circuit solution to equalize currents in the assembled transistors. Traditionally, to equalize the currents of parallel connected transistors in their emitter circuits set the equalizing resistors, voltage drop on which, depending on the dispersion of the gain, should be in the range from 0.2 to 0.7 V. If gain variation is not more than *30–50%*, it is enough to create voltage drop on equalizing resistors equal to *25%* of base-emitter voltage of output transistor *VT6*. That is, with an operating base-emitter voltage of 1 V the voltage drop across the equalizing resistor should be 0.25 V.

To calculate the ratings of equalizing resistors, let's calculate the maximum collector current of one transistor of the composite output stage. At total load current of 50 A, passing through the composite output stage of five transistors, the current of one transistor will be five times less, i.e. 10 A.

Let's calculate ratings of equalizing resistors *R19-R23:*

$$R_{19} = \frac{\Delta U_{R19}}{I_{LOAD.VT6}} = \frac{0.25}{10} = 0.025 \, \text{ohm} \tag{4}$$

where:—voltage at *R19; I_{LOAD}*.—is the maximum load current of *VT6*.

At maximum operating temperature the increase of collector return currents can lead to spontaneous opening of the transistors of the output stage. Therefore it is necessary to install a protective resistor *R17* between bases of output transistors and positive output busbar of the regulator.

The rating of resistance *R17* is chosen so that the collector reverse currents flowing through it do not create an opening voltage drop between the base and emitter, which according to the input characteristics of 2SC5200 transistor is $U_{BEVT6} = 0,4$ V. We set the voltage drop equal to *30%* of the opening voltage between the base and emitter of the output transistor, that is 0.12 V.

To calculate the nominal resistance *R17* we calculate the collector return current of one of the compound transistors at maximum operating temperature:

$$I_{C.R.VT6} = I_{C.R.VT6(20)} \times e^{(0.1...0.13)(t_{CMAX}-20)}$$
$$= 5 \times 10^{-6} \times e^{(0.1...0.13)\times(50-20)} = 200 \times 10^{-6} \, \text{A} \tag{5}$$

where: $I_{C.R.VT6}$—reverse collector current of *VT6; $t_{C.MAH} = 50°$* C.

Let's determine the total reverse collector current of the composite output cascade:

$$I_{C.T.C.} = 200 \times 10^{-6} \times 5 = 1 \times 10^{-3} \, \text{A} \tag{6}$$

where: $I_{C.T.C.}$—is the total reverse current of collectors of the compound cascade.

Let's calculate the nominal resistance *R17:*

$$R_{17} = \frac{U_{BE.VT.MIN.}}{I_{C.T.C.}} \times 0.3 = \frac{0.4}{1 \times 10^{-3}} \times 0.3 = 120\,\text{ohm} \qquad (7)$$

where: UBE.VT6.MIN —is the minimum base-emitter opening voltage of VT6.

To select the pre-terminal stage transistor, let's calculate the total base current of the composite output stage:

$$I_{B.VT6-VT11} = \frac{I_{L.T.}}{\beta_{VT6}} = \frac{50}{60} = 0.833\,\text{A} \qquad (8)$$

where: $I_{L.T.}$—maximum load current; β_{VT6}— current amplification factor of VT6.

The total base current of the compound transistor is 0.833 A which flows through the collector-emitter junction of the pre-terminal stage. Therefore we choose 2SC5171 transistor as a pre-terminal stage, which has maximum collector current of 2 A and maximum power dissipation 20 W.

Calculation of *R9* resistance is done at minimum load current and maximum ambient operating temperature. The calculation is made under the condition that the voltage drop on *R9,* created by the collector return current, does not lead to the opening of the base-emitter junction of the transistor.

To calculate the rating of resistance *R9* it is necessary to calculate the value of the reverse current of the transistor of the pre-terminal stage:

$$I_{C.R.VT4} = I_{C.R(20°C)} \times e^{(0.08...0.13)(t_{MAX}-20)}$$
$$= 1 \times 10^{-6} \times e^{0.1 \times (50-80)} = 20 \times 10^{-6}\,\text{A} \qquad (9)$$

where: $I_{C.R.VT4}$—is reverse collector current of *VT4;* $I_{C.R(20)}$—is reverse collector current of *VT4* at 20 °C; $t_{CMAH} = °C$.

Let's set voltage drop across resistor *R9* as *30%* of total breakaway voltage of base-emitter transistors of output and pre-terminal stages, which is *0.40 + 0.46 =* 0.86 V.

Then the value of resistance *R9* can be determined from the expression:

$$R_9 = \frac{U_{BE.VT4,VT6}}{I_{C.R.VT4,VT6}} \times 0.3 = \frac{0.86}{0.20 \times 10^{-3}} \times 0.3 = 1290\,\text{ohm} \qquad (10)$$

where: $U_{BE.VT4.VT6}$—total base-emitter voltage of VT4 and VT6.

From the series *E24* we choose the nominal resistance *R9* equal to *1.3 kOhm.*

Through the resistance *R4* is formed base current of the pre-terminal stage and collector current of the control circuit matching stage. Therefore to calculate the rating of resistance *R4* it is necessary to calculate the current through resistor *R4* at minimum supply voltage and maximum load current.

To calculate the rating of resistance *R4* it is necessary to calculate the voltage drop across resistance *R4* and the amount of current flowing through R4.

The rating of *R4* can be determined from the expression:

$$R_4 = \frac{\Delta U_{R4}}{I_{R4}} \tag{11}$$

where: ΔU_{R4}– minimum voltage drop across R4 at minimum input supply voltage; I_{R4}—current through R4.

The current through resistance R4 can be determined from the expression:

$$I_{R4} = I_{BVT4} + I_{R9} + I_{C.T.R.VT3} \tag{12}$$

where: I_{R4}—current through R4; I_{BVT4}—base current of transistor VT4; I_{R9}—current through R9; $I_{C.T.R.VT3}$—technological collector current of transistor VT3 (20% of current I_{R4}).

The minimum voltage drop across R4 is found from the expression:

$$\Delta U_{R4} = E_1 + E_{2MIN} - U_{OUT.R.V.} - \Delta U_{R9}$$
$$= 3 + 13.6 - 12 - 2.2 = 24 \text{ V} \tag{13}$$

where: ΔU_{R4}– voltage on R4 at the lowest supply voltage; E_1—surge voltage; E_2—the lowest supply voltage; $U_{OUT.R.V.} = 12$ V; $\Delta U_{R9} = 2.2$ V.

To calculate the current through resistance R9, let's determine the voltage drop across R9:

$$\Delta U_{R9} = \Delta U_{R19} + U_{BE.VT6} + U_{BE.VT4} = 0.25 + 1 + 0.95 = 2.2 \text{ V} \tag{14}$$

where: ΔU_{R9}– voltage drop across R9; $U_{BE.VT6} = 1$ V—base-emitter voltage VT6; $U_{BE.VT4} = 0.95$ V—base-emitter voltage VT4.

Let's determine the current through resistance R9:

$$I_{R9} = \frac{\Delta U_{R9}}{R_9} = \frac{2.2}{1300} = 0.00169 \text{ A} \tag{15}$$

The voltage drop across resistance R17 is determined from the expression:

$$\Delta U_{R17} = \Delta U_{R19} + U_{BE.VT6} = 0.25 + 1 = 1.25 \text{ V} \tag{16}$$

where: ΔU_{R17}, ΔU_{R19}– voltages on resistors R17, and R19, respectively; $U_{BE.VT6}$—voltage of the base-emitter junction of VT6.

Now we can calculate the current flowing through the resistance R17:

$$I_{R17} = \frac{U_{R17}}{R_{17}} = \frac{1.25}{120} = 0.0104 \text{ A} \tag{17}$$

where: I_{R17}—is the current flowing through resistor R17.

To determine the emitter current of transistor VT4 it is necessary to calculate the collector current of transistor VT5 voltage divider resistors R13-R15.

We will determine the collector current of *VT5* from the expression:

$$I_{C.VT5} = \frac{U_{OUT.R.V.} + \Delta U_{R17}}{R18} = \frac{12 + 1.25}{43000} = 0.000308 \text{ A} \qquad (18)$$

The voltage divider current of resistors *R13–R15* will be determined from the expression:

$$I_{R13-R15} = \frac{\Delta U_{R17}}{R_{13} + R_{14} + R_{15}} = \frac{1.25}{2000 + 3000 + 1000} = 0.000208 \text{ A} \qquad (19)$$

The emitter current of transistor *VT4* is determined from the expression:

$$\begin{aligned} I_{E.VT4} &= I_{CVT5} + I_{R13-R15} + I_{R17} + I_{B.VT6-VT11} \\ &= 0.000308 + 0.000208 + 0.0104 + 0.833 = 0.84391 \text{ A} \end{aligned} \qquad (20)$$

where: $I_{E.VT4}$—emitter current of *VT4*;

Find the collector current of transistor *VT4* from the expression:

$$I_{C.VT4} = \frac{I_{E.VT4.}}{1 + \frac{1}{\beta_{VT4}}} = \frac{0.84391}{1 + \frac{1}{100}} = 0.8355 \text{ A} \qquad (21)$$

where: $I_{C.VT4}$—collector current *VT4*; $I_{E.VT4}$—emitter current *VT4*;—current transfer coefficient *VT4*.

Let's calculate the base current of *VT4* pre-terminal stage:

$$I_{B.VT4.IN.} = \frac{I_{C.VT4.}}{\beta_{VT4}} = \frac{0.8355}{100} = 8.355 \times 10^{-3} \text{ A} \qquad (22)$$

where: $I_{B.VT4\ MIN.}$—is the *VT4* base current at maximum load and minimum supply voltage; $I_{C.VT4}$—is the *VT4* collector current.

Now according to expression (12) we can calculate the current of resistor *R4*:

$$I_{R4} = I_{BVT4} + I_{R9} + I_{C.T.R.VT3} = (0.008355 + 0.00169) \times 0.2 + 0.01004 = 0.01204 \text{ A}$$

Now according to expression (10) the nominal *R4* can be determined from the expression:

$$R_4 = \frac{\Delta U_{R4}}{I_{R4}} = \frac{2.4}{0.01204} = 199.3 \text{ Ohm.}$$

From series *E 24* we choose the nominal *R4* of 200 Ω.

Determine the power dissipation at the output transistors and the pre-terminal stage.

To calculate the maximum power dissipation at the pre-terminal stage we calculate the maximum voltage drop across the collector-emitter junction of the output regulating transistor *VT4:*

$$\Delta U_{CE.VT4.} = E_1 + E_{2MAX.} - U_{OUT.R.V.} - \Delta U_{R17} =$$
$$= 3 + 20.4 - 12 - 1.25 = 10.15V$$
(23)

where: $\Delta U_{CE.VT4MAX.}$– voltage drop at collector-emitter junction of transistor *VT4*; E_1—surge voltage; $V_{2MAH.}$ maximum input supply voltage of output regulating stage; $U_{OUT.R.V.} = 12$ *V*; ΔU_{R19}– voltage drop across *R19*.

Power dissipation on the pre-terminal stage we will determine from expression:

$$P_{OUT.VT4} = \Delta U_{CE.VT4MAX.} \times I_{CVT4} = 10.15 \times 0.8355 = 8.4803 \text{ W}$$
(24)

where: $P_{OUT.VT4}$—power dissipated at 2SC5171 *(VT4)*; $\Delta U_{CE.VT4MAX.}$– maximum voltage between collector and emitter of pre-terminal stage; $I_{C.VT4}$—maximum collector current of—*VT4*.

To calculate the maximum power dissipation on one of the output transistors, it is necessary to calculate the maximum voltage drop across the output transistor:

$$\Delta U_{CE.VT6.MAX.} = U_{InPUT.MAX.} - U_{OUT.R.V.} - \Delta U_{R19} = 20.4 - 12 - 0.25 = 8.15 \text{ V}$$
(25)

where: $U_{CE.VT6.MAX.}$—collector-emitter voltage drop of *VT6*; $U_{InPUT.MAX.}$—is the maximum input voltage E_2; $U_{OUT.R.V.} = 12$ *B*.

Let's calculate the maximum power dissipation at one of the five transistors of the compound stage *(VT5):*

$$P_{OUT.VT6} = \Delta U_{CE.VT6MAX} \times \frac{I_{LOAD.MAX.}}{5} - I_{B.VT6} = 8.15 \times \frac{50}{5} - 0.1666 = 80.14 \text{ W}$$
(26)

where: $P_{OUT.VT6}$—power dissipated at 2SC5200; $\Delta U_{CE.VT6MAX.}$– voltage drop at collector-emitter junction of transistor *VT6*; $I_{NAG.MAX.}$—maximum load current.

In any circuit design solution voltage regulator control circuit is the basis of accuracy stabilization is the stability of the voltage node reference voltage.

Stabilitrons produced today have a finite internal resistance and therefore the stability of their voltage always depends on the current flowing through the stabilizer. In the proposed schematic solution the main reference gate *VD2* is fed from the current regulator on the transistor *VT2*. In this case, the stabilizer *VD1* current regulator on the transistor *VT2* is powered from the stabilizer on the transistor *VD1*. With this circuit solution a stable current will flow through the precision transistor.

Apart from the reference voltage the stability of the output voltage of the regulator depends on the changes of the load current. In suggested circuit solution the increase of stability of output voltage with load current changes is solved by the introduction of positive current feedback, which signal is formed on resistance *R7*. When increasing the load current the voltage increment on the feedback resistance *R7* is added to the voltage of the reference voltage, thus correcting the decrease of output voltage with increasing load current.

Stabilitrons *1N5235B* are used as main and auxiliary voltage regulators *VD1* and *VD2*. Opposite current regulators are made on a complementary pair of transistors *2N5551B* and *2N5401*.

The values of rated currents of *VD1* and *VD2* stabilizing diodes are set by the ratings of resistors *R2* and *R3*.

The rating of resistors *R2* and *R3* can be determined from the expression:

$$R_{2,3} = \frac{U_{VD1} - \Delta U_{BE.VT1,VT2}}{I_{.VD1}} = \frac{6.8 - 0.84}{0.02} = 298 \, \text{Ohm} \tag{27}$$

where: U_{VD1}—nominal stabilization voltage of *VD1*; $\Delta U_{BE.VT1,VT2}$— base-emitter voltage of transistors *VT1* and *VT2* according to the input characteristics is *0,84 V*; I_{VD1}—nominal stabilization current of *VD1*.

From the series of *E24* nominal values of *R2* and *R3* will be equal to *300 Ω*.

In order to ensure a stable start of the voltage reference node at the moment of supply voltage, resistor *R1* is included in the circuit of the support node equal to *100 kOhm*.

To determine the power dissipation on the current stabilizer transistors, let's determine the voltage drop across the resistors *R2* and *R3*:

$$\Delta U_{R2,R3} = \Delta U_{.VD1} - \Delta U_{BE.VT,VT2} = 6.8 - 0.7 = 6.1 \, \text{V} \tag{28}$$

where: voltage drop across *R3;*—nominal stabilization voltage of *VD1* and *VD2;*— base-emitter voltage of transistors *VT1* and *VT2*.

The value of the current through resistor R3 is determined from the expression:

$$I_{R3} = \frac{\Delta U_{R3}}{R_3} = \frac{6.1}{300} = 20.3 \times 10^{-3} \, \text{A} \tag{29}$$

where: I_{R3}—current *R3*; ΔU_{R3}− voltage drop on *R3*.

The voltage drops at collector-emitter junctions *VT1* and *VT2* will be equal:

$$\Delta U_{CE.VT1,VT2} = E_1 + E_{2MAKC,} - \Delta U_{R3} - U_{VD1,VD2} = 20.4 + 3 - 6 - 6.8 = 10.6V \tag{30}$$

where: $\Delta U_{CE.VT1,VT2}$− collector-emitter voltages of *VT1* and *VT2*; E_{2MAX}— maximum input voltage; E_1—surge voltage; ΔU_{R3}− voltage drop across resistor *R3*; $U_{VD1,VD2}$—nominal stabilization voltage of stabilizers *VD1* and *VD2*.

Now we can determine the power dissipated by transistors *VT1* and *VT2*:

$$P_{DIS.VT1,VT2} = I_{C.VT1,VT2} \times \Delta U_{CE.VT1,VT2} = 20 \times 10^{-3} \times 10.6 = 0.212 \, \text{W} \tag{31}$$

where: $P_{DIS.VT1mVT2}$—power dissipated on $VT1$ and $VT2$; $I_{CVT1,VT2}$—collector current of $VT1$ and $VT2$; $\Delta U_{CE.VT1,VT2}$— collector-emitter voltage of transistors $VT1$, $VT2$.

Calculated value of power dissipation at current stabilizing transistors 0,212 W is three times less than maximum allowable power 0.625 W.

The node of matching of control signals with the pre-terminal stage in this circuit design is made on transistor $VT3$ and operational amplifier $AD1$.

To choose the type of transistor $VT3$ we will determine the maximum power dissipation of the cascade.

Maximum power will be dissipated at transistor $VT3$ at maximum input supply voltage E_2 and minimum load current.

Determine the maximum voltage drop across resistance $R4$:

$$\Delta U_{MAX.R4} = E_{2MAX} + E_1 - U_{OUT.R.V.} - \Delta U_{BE.VT6}$$
$$- \Delta U_{BE.VT4} = 20.4 + 3 - 12 - 0.95 - 1 = 9.45 \, \text{B} \tag{32}$$

where: E_{2MAX}—maximum input voltage; E_1—surge voltage; $U_{OUT.R.V.}$—rated output voltage of the stabilizer; $\Delta U_{BE.VT4}$ and $\Delta U_{BE.VT6}$—base-emitter voltage of transistors $VT4$ and $VT6$ at the time of opening.

Let's calculate the maximum current through resistor $R4$:

$$I_{R4MAX} = \frac{\Delta U_{MAX.R4}}{R_4} = \frac{9.45}{200} = 0.0472 \, \text{A} \tag{33}$$

Determine the maximum collector current of $VT3$:

$$I_{C.VT3} = I_{R4MAX} - I_{R9} = 0.0472 - 1.69 \times 10^{-3} = 45.51 \times 10^{-3} \, \text{A} \tag{34}$$

where: $I_{C.VT3}$—collector current of $VT3$; I_{R9}—current of $R9$.

In order to reduce the power dissipation on the transistor $VT3$, a resistance $R5$ of $100 \, ohms$ is introduced into the circuit.

The voltage drop across resistance $R5$ is determined from the expression:

$$\Delta U_{R5} = I_{C.VT3} \times R_5 = 45.51 \times 10^{-3} \times 100 = 4.551 \, V \tag{35}$$

The maximum voltage drop across the collector-emitter junction of $VT3$ is determined from the expression:

$$\Delta U_{CE.VT3} = E_{2MAX} + E_1 - \Delta U_{R4} - \Delta U_{R5} - \Delta U_{R6}$$
$$= 20.4 + 3 - 9.68 - 4.551 - 4.551 = 4.618 \, V \tag{36}$$

where: $\Delta U_{CE.VT3}$— voltages at collector-emitter of $VT3$; E_{2MAX}—maximum input voltage; E_1—surge voltage; ΔU_{R4}, ΔU_{R5} and ΔU_{R6}— voltages at $R4$, $R5$ and $R6$ respectively.

Let's calculate the power dissipation at transistor $VT3$:

$$P_{DIS.VT3} = I_{C.VT3} \times \Delta U_{CE.VT3} = 0.04551 \times 4.618 = 0.21016\,W \qquad (37)$$

where: $P_{DIS.VT3}$—power dissipation at VT3; I_{KVT3}—collector current of VT3;—collector-emitter voltage of VT3.

To prevent spontaneous opening of transistor VT3 when the collector return current increases at maximum operating temperature, a 100 kOhm resistor is placed in the base circuit of VT3.

The operating mode of the operational amplifier is set by the ratio of the resistors R10 and R12, which determine the gain of the operational amplifier and, accordingly, the stabilization coefficient of the whole enviromental support system. The ratings of resistances R10, R12 and the rating of the correcting capacitor C1 are selected when debugging the system.

The output voltage divider on resistors R26-R29 forms the negative feedback signal from potentiometer R27. A thermistor with a positive temperature coefficient of resistance (+TKR) is included in the lower arm of the voltage divider.

Calculation of output voltage divider parameters is done as follows:

We will set the output voltage divider current to 0,004–0,007% of the maximum load current of 50 A. So the current of the divider should be 0,022–0,035 A.

Let's calculate the total resistance of the output voltage divider:

$$R_{V.D.} = \frac{U_{OUT.R.V.}}{I_{V.D.}} = \frac{12}{0.0225} = 535\,Ohm \qquad (38)$$

where: $R_{V.D.}$—total resistance of resistors R26-R29; $U_{OUT.R.V.}$—rated output voltage; $I_{V.D}$—total current of output voltage divider.

Then the resistance of the output voltage divider is R26 = 150 ohms, R27 = 180 ohms, R28 = 180 ohms, R29 = 27 ohms.

In order to increase the thermal stability of the output voltage regulator thermistor R29 is selected when debugging the system when changing the ambient temperature from 27 °C to 50 °C.

Dynamic load node on the transistors VT5, VT12 and VT13 connects additional load R25 when the load current is less than 0,3 A. The threshold of additional load R25 is chosen by adjusting the potentiometer R14.

Results of simulation of the combined output stage and pre-terminal stage are presented in Table 1.

The results of modeling of the matching cascade are presented in Table 2.

A comparative analysis of calculations and modeling results is presented in Table 3.

As the simulation results have shown, the above calculations with sufficient accuracy correspond to the simulation results.

Table 1 Results of simulation of the termination, pre-terminal and matching cascades

Parameters	При $U_{InPUT.MAX}$ и $I_{LOAD} = 50\,A$	При $U_{InPUT.R.V}$ и $I_{LOAD} = 50\,A$	При $U_{InPUT.MIN}$ и $I_{LOAD} = 50\,A$
E1 + E2	23.39967 V	19.99967 V	16.599967 V
E2	20.39967 V	16.99967 V	13.599967 V
$U_{OUT.R.V}$	12.00044 V	12.00001 V	11.99956 V
U_{VD2}	6.80029 V	6.80027 V	6.80024 V
I_{LOAD}	50.00184 A	50.00003	49.99817 A
I_{BBT6}	0.1667055 A	0.16669951 A	0.16669333 A
I_{ET6}	10.00233 A	10.00197 A	10.00160 A
I_{CT6}	9.83563 A	9.83527 A	9.83491 A
U_{BVT6}	13.2625 V	13.26205 V	13.26160 V
U_{EVT6}	12.2505 V	12.25006 V	12.24960 V
P_{VT6}	80.32086 W	46.88239 W	13.44644 W
I_{BVT4}	8.36302 mA	8.36272 mA	8.36242 mA
I_{EVT4}	0.84466529 A	0.84463508 A	0.84460409 A
I_{CVT4}	0.83630227A	0.83627235 A	0.83624167 A
U_{BVT4}	14.21075 V	14.21030 V	14.20985 V
U_{EVT4}	13.2625 V	13.26205 V	13.2616 V
P_{VT4}	8.48567 W	5.64241 W	2.79936 W
I_{BVT3}	240.75828 mkA	123.96546 mkA	14.26731 mkA
I_{EVT3}	36.12217 mA	19.00784 mA	1.90076 mA
I_{CVT3}	35.88141 mA	18.88388 mA	1.88649 mA
U_{BVT3}	4.39892 V	2.66774 V	0.89440898 V
U_{EVT3}	3.61221 V	1.90078 V	0.19007383 V
U_{CVT3}	10.62261 V	12.32191 V	14.0212 V
P_{VT3}	0.2517325 W	0.19688642 W	0.02610236 W

4   Debatable Issues

The question of approaches to designing information systems for technical universities remains open. Most developers consider a training system only as a technical device, which is the main reason for the low didactic efficiency of existing software and hardware educational resources.

New technical solutions adopted in the design cannot be the main criteria for the quality of the work performed. In our opinion, the design of an information learning system can be considered successful only if the system fulfills the didactic tasks of education.

New technical solutions can only confirm the correctness of the implementation of the algorithms embedded in the system. But if the information system does not take

Table 2 Results of simulation of the reference voltage node

Parameters	При $U_{InPUT.MAX}$ и $I_{LOAD} = 50$ A	При $U_{InPUT.R.V}$ и $I_{LOAD} = 50$ A	При $U_{InPUT.MIN}$ и $I_{LOAD} = 50$ A
I_{BVT1}	130.71799 mkA	134.7138 mkA	130.97445 mkA
I_{EVT1}	20.10704 mA	20.104 mA	20.10086 mA
I_{CVT1}	19.97632 mA	19.96929 mA	19.96189 mA
U_{BVT1}	6.80029 V	6.80027 V	6.80024 V
U_{EVT1}	6.03178V	6.03087 V	6.02993 V
U_{CVT1}	16.5994 V	13.19948 V	9.79957 V
P_{VT1}	0.21120261 W	0.1432558 W	0.075356 W
I_{BVT2}	163.47567 mkA	169.00146 mkA	174.91630 mkA
I_{EVT2}	20.28545 mA	20.28206 mA	20.27856 mA
I_{CVT2}	20.12197 mA	20.11306 mA	20.10364 mA
U_{BVT2}	16.5994 V	13.19948 V	19.79957 V
U_{EVT2}	17.31403 V	13.91505 V	10.5161 V
U_{CVT2}	6.80029 V	6.03087 V	6.80024 V
P_{VT2}	0.21167403 W	0.14322097 W	0.07482762 W
U_{R7} +	333.38438 mkB	333.25905 mkB	333.13342 mkB
U_{R7} -	0	0	0
I_{R25}	0.0412168 mkB	0.04121539 mkB	0.0412138 mkB

Table 3 Comparative data of calculations and modeling

Parameters	Calculated	Simulated
I_{R3}	20 mA	20.28544 mA
I_{R4}	12.04 mA	11.949 mA
I_{R4MAX}	47.2 mA	45.944 mA
I_{R9}	1.69 mA	1.7022 mA
I_{R17}	10.4 mA	10.51 mA
I_{CVT3}	45.51 mA	46.903 mA
I_{CVT4}	0.8355 A	0.83627 A
I_{EVT4}	0.84391 A	0.844632 A
I_{BVT4}	8.355 mA	8.3627 mA
$U_{CE.VT1-VT2}$	10.6 V	10.538 V
$P_{DIS.VT1-VT2}$	0.212 W	0.2112 W
$P_{DIS.VT3}$	0.21016 W	0.20258 W
$P_{DIS.VT4}$	8.4803 W	8.48564 W
$P_{DIS.VT6}$	80.14 W	80.32065 W

into account pedagogical concepts in education and does not contain the algorithms of didactically effective teaching methods, the new technical solutions will not help to improve the effectiveness of the educational process.

Today there is an opinion that only students need to be trained and educated. It is well known that not all university teachers are harmoniously nurtured individuals. That is why the issue of andragogy, i.e. educating teachers, is so acute today. Cognitive motivation, interest to learning and effectiveness of mastering educational material directly depend on the inner state of a teacher, the negative manifestation of which makes it impossible to create a harmonious psycho-emotional working environment.

5 Conclusions

Functional capabilities of technical educational tools should be focused on didactic tasks firstly and on the subject area secondly.

Informational teaching systems should be focused on the application of student-centered approach in education, which takes into account the personal characteristics of all participants in the educational process, both students and teachers.

In order to identify the personal characteristics of the participants of the educational process, the information learning system should contain automated means of psychodiagnosis.

The use of psycho-factors as additional elements of system analysis significantly increases the effectiveness of the analysis of the functioning and quality of the educational process.

Taking into account the non-linearity and unpredictability of the pedagogical process, the information learning system should be able to promptly adjust the teacher's work algorithm and the form of teaching material presentation.

References

1. Polyankina, S.Yu.: Semiotic approach to the resolution of the key contradictions of the modern education system. Bullet. Tomsk State Univ. Philos. Soc. Polit. Sci. 41, 64–71 (2018)
2. Chernikova, V.E.: Spirituality in the modern educational process. Scientific Prob. Humanitarian Res. 6, 135–142 (2010)
3. Andreev, V.V., Gorbunov, V.I., Evdokimova, O.K., Rimondi, G.: Transdisciplinary approach to improving study motivation among university students of engineering specialties. Educ. Self Dev. 15(1), 21–37 (2020)
4. Campbell, R., Soenens, B., Beyers, W., et al.: University students' sleep during an exam period: the role of basic psychological needs and stress. Motiv. Emot. 42, 671–681 (2018). https://doi.org/10.1007/s11031-018-9699-x
5. Erten, I.H.: Interaction between Academic Motivation and Student Teachers' Academic Achievement. Procedia Soc. Behav. Sci. 152, 173–178. (2014). https://doi.org/10.1016/j.sbspro.2014.09.176

6. Avakyan, I.B.: On the relationship between the innovative readiness of teachers and the socio-psychological climate of universities. Educ. Sci. **20**(4), 114–131 (2018). https://doi.org/10.17853/1994-5639-2018-4-114-131

7. Avakyan, I.B.: Striving for self-development as a factor of innovative readiness of university teachers. Educ. Self-Dev. **15**(2), 88–131 (2020). https://doi.org/10.26907/esd15.2.08

8. Glushan, V.M.: Construction of computer learning systems with adaptation to the psycho-emotional state of the learner. Bullet. Taganrog State Pedagogical Inst. **1**, 178–184 (2008)

9. Glushan, V.M., Izvestia, S.F.U.: Computer technologies and problems of building automated training and controlling systems. Techn. Sci. **7**(144), 237–242 (2013)

10. Lagunova, M.V., Yurchenko, T.V. Management of cognitive activity of students in the information and educational environment of the university: monograph. N. Novgorod: NNGASU, 167 p. (2011)

11. Shikov, A.N.: On the role of the teacher in the application of computer learning technologies. Proc. Sociosphere Res. Center **1**, 126–132 (2013)

12. Pechnikov, D.A., Pechnikova, L.G.: Psychological and pedagogical approach to the creation of computer learning technologies. Fundamental Appl. Res. Modern World **18**(1), 11–20 (2017)

13. Gordienko, S.A., Trubachev, I.V., Yarmysh, V.A.: Using expert system to calculate the indicator of theoretical training of the student. Innovative Technologies in Educational Process: Materials of XIX All-Russian Scientific and Practical Conference, pp. 55–60. Moscow (2017)

14. Khramov, B.V., Vitchenko, O.V., Tkachuk, E.O., Golubenko, E.V.: Intelligent methods, models and algorithms of educational process organization in modern universities. Rostov n/D: FGBOU VO RSUPS, 152 (2016)

15. Tsybov, N.N.: Cognitive automated learning system. Patent no. 2229 Kyrgyz Republic, IPC G09B 19/00, G09B 9/048. No. 20190079.1; Application. 11.11.2019; Republ. 30.11.20, Intellectualdyk menchik rasmiy Bulletin 11(259), 2 p.: ill. (2020)

16. Tsybov, N.N.: A method of increasing the efficiency of perception, processing and assimilation of new learned information. Patent for invention KG 2303 Kyrgyz Republic, MRK G09B 5/14. No. 20220006.1; Application. 27.01.2022; Republ. 30.08.22, Intellectualdyk menchik rasmyy bulletin 8, 6 p. (2022)

17. Tsybov, N.N.: Precision thermostable DC voltage stabilizer with compensation of internal resistance. Patent for invention KG 2029 Kyrgyz Republic, IPC G05F 1/56. No. 20170075.1; Application. 20.06.17; Republ. 28.02.18, Intellectualdyk menchik rasmii bul. 2, 2 p.: ill (2018)

An Online Course as the Basis for Training of Modern Volunteers

Valeriia Sergeevna Sirenkoⓘ **and Aleksandra Zelko**ⓘ

Abstract The chapter offers a new solution for the training of volunteers—the online course "Fundamentals of volunteering". This course was developed by the authors and has been implemented on the basis of Immanuel Kant Baltic Federal University (Russia, Kaliningrad) since 2020, has been implemented on the lms.kantiana platform since 2021. The course is available to students of all forms and levels of education. This course is designed for those who are interested in volunteering and intend to start a volunteer practice. The course program is aimed at the formation of competencies in accordance with the labor functions of a teacher, facilitator, educator, or social worker. The purpose of the course "Fundamentals of volunteering" is to form professional competence in the field of organizing work with youth by mastering the basic knowledge of the specifics of volunteering by students, taking into account the experience of implementing practices in the field of volunteering. The program of the online course is described, examples of tasks for future volunteers and an assessment system are given. The significance and timeliness of the implementation of this online course is discussed: Hosting the 2018 FIFA World Cup in Russia, The COVID-19 pandemic, significant government measures to support volunteers, and etc.

Keywords Online course · Training · Volunteers · Voluntary service · Supplementary vocational education

V. S. Sirenko (✉)
Immanuel Kant Baltic Federal University, Kaliningrad 236041, Russian Federation
e-mail: VSirenko@kantiana.ru

A. Zelko
Los Angeles City College, Los Angeles, CA 90029, USA

Z. Dvořáková and A. Kulachinskaya (eds.), *Digital Transformation: What is the Impact on Workers Today?*, Lecture Notes in Networks and Systems 827,
https://doi.org/10.1007/978-3-031-47694-5_7

1 The Introduction

The pandemic of the new coronavirus infection is undoubtedly one of the most significant shocks of recent decades. Besides an immediate "biological" impact on the human body, the COVID-19 pandemic and widespread anti epidemic measures which have been introduced have fundamentally affected the mode of life of many people [1].

Due to the COVID-19 epidemic, universities have adopted online teaching to allow students to study courses online. The internet is a common learning platform for learners and teachers to interact, communicate, and collaborate in a specific way [2, 3], and the use of information technology (IT) in teaching has been implemented worldwide.

The purpose of developing online learning is to use IT to enhance the quality of teaching and learning, creating a high-quality learning environment, eliminating time and space constraints on learning, improving the management of teaching resources, and establishing the integration of IT with teaching and learning in various fields [3].

Online learning is the use of a wide range of technologies such as the internet, email, chat, new groups and texts, audio and video conferencing transmitted across electronic networks to transfer education [2].

2 The Literature Review

Many educational institutions have started to develop and use online training for their volunteers. The online volunteer training course has become one of the most popular online courses at Immanuel Kant Baltic Federal University (IKBFU). There are several reasons:

1. Hosting the 2018 FIFA World Cup in Russia.
   The International Olympic Committee has recognized volunteers as the main participants in the process of organizing sports events [4]. This makes volunteers the most important stakeholders of the event [5].
2. The COVID-19 pandemic has led to the fact that the need for volunteers has increased several hundred times. We have already noted in our previous study that one of the youngest and most relevant areas of volunteering is online one, which is gaining increasing mass and popularity in connection with the modern realities of the global pandemic and at the same time with a fairly high level of development of information and communication technologies [6].
3. The state began to take significant measures to support volunteers. Federal Act No. 135 of 11 August 1995 On charitable activities and volunteering [7] was amended (08.12.2020). 2018 has been declared the year of the volunteer in Russia. The International Volunteer Day began to be celebrated on December 5th. The number of non-profit organizations and youth volunteer associations has increased.

Abramova S. V. notes that volunteering has become significant both for society and the state, and for the volunteers themselves. Voluntary service helps the state to more effectively solve the problems. The development of volunteering contributes to the formation of social activity of citizens, the formation of civil society; it positively affects the social and economic development of the country, helping to solve socially significant problems [8].

4. Wide experience in practical activities led to the understanding that volunteers should not only be motivated and supported, but also prepared for the service [9]. Volunteers should be familiar with the historical, legal and ethical foundations of volunteering. Also they should comprehend the psychological and pedagogical changes that occur to them during voluntary service, and they should be able to interact with other people.

5. The university has developed a system of benefits and opportunities for practicing volunteers. This is an extremely important remark, since it is crucial to create a personnel reserve of employees with a practical understanding of professional activity, it is necessary to develop targeted forms of support for student volunteers [10]

6. Availability of an online platform and a course, developed by professionals, at university. This allows young people from anywhere to be trained, immediately receive a certificate and have the opportunity to become a volunteer after training.

Frendo [11] analyzed three different modes of delivery for volunteer training (online vs. face- to- face vs. blended) and found that although face-to-face training was highly regarded in creating connections with other volunteers, online training met volunteers' motivational needs when opportunities were provided for learners to interact in real-time with people from a distance.

The benefits to online volunteer training may outweigh the barriers. Although technical difficulties and limited interaction amongst participants can create barriers during training, this is offset by tangible benefits such as the ability to draw participants from a larger geographic region, provide flexible training times, and engage participants in cross cultural learning [12, 13].

Online volunteering will continue to increase for its cost-effectiveness but at the expense of time and operational cost in adopting ITs [14].

Some researchers note that most of the surveys in the field of volunteering are aimed at studying the personality of the student, and not at the student environment of the university. For example, these include the educational courses offered by the university, the strategic mission and priorities of the university, or the history of student associations that exist in the educational institution [15].

This critically limits the number of studying youth and opportunities for voluntary service. Therefore, the authors were looking for a special way to prepare young people for volunteering.

3 Research Methodology and Design

The online course of supplementary vocational education "Fundamentals of volunteering" has been implemented on the lms.kantiana platform since 2021. The course is available to students of all forms and levels of education. Upon completion of training and successful completion of the course, the student is handed a certificate of achievement. This course is designed for those who are interested in volunteering and intend to start a volunteer practice. The course program is aimed at the formation of competencies in accordance with the labor functions of a teacher, facilitator, educator, or social worker.

The purpose of the course "Fundamentals of volunteering" is to form professional competence in the field of organizing work with youth by mastering the basic knowledge of the specifics of volunteering by students, taking into account the experience of implementing practices in the field of volunteering.

As a result of the training, the graduate of the program will be able to:

- engage in social interaction and actualize their role in the team;
- organize an independent interaction with representatives of social services, volunteer organizations;
- implement practices in the field of volunteering;
- conduct an analysis of their own volunteer activities.

The length of the course is academic 36 h. It is designed for 5 weeks (testing is conducted at the end of each topic of the course). The course consists of short (5–10 min) video lectures, a variety of information materials (presentations, educational texts, fragments of documents), tests on each topic of video lectures, summative assessment.

The course "Fundamentals of volunteering" is an element of the system for preparing volunteers for social service.

Within the course, attention is paid to such issues as the concept of volunteering, the goals and objectives (basic and applied) of volunteering, the history of the development of volunteerism in Russia, European countries and the USA, the state policy of the Russian Federation in the social sphere and in particular in the field of volunteerism, the legal framework of volunteer activity, the elements of the mechanism for the practical implementation of volunteer activity.

The online course "Fundamentals of volunteering" has the following structure (see Table 1).

The course is aimed at developing the following universal competencies (UC) and general professional competencies (GPC) (see Table 2).

At the moment 259 students are enrolled in the course or have completed it. 57 students have completed the program (8 students passed the summative assessment with less than 30 points, 20 students received from 30 to 49 points, 37 students have completed the summative assessment with 50 points and more (maximum score—65) (Table 3).

Volunteer orientation and training also includes [18]:

Table 1 The online course "Fundamentals of volunteering" structure

Module (course section)	Lesson (topic within a section)	Educational material (video, presentations, texts)	Control tasks (tests, tasks for self-examination, etc.)
1. Fundamentals of volunteering	1.1 The history of the volunteer movement in Russia and abroad	The history of the volunteer movement (presentation and video), Universal Declaration on Volunteering, 2001 [16], Federal Act No. 135 of 11 August 1995 On charitable activities and volunteering [7]	Test
	1.2 Basic terms and general approaches to volunteering	Basic terms and general approaches to volunteering (presentation and video), Concept development of volunteering in the Russian Federation up to 2025 [17]	Test
	1.3 Legal basis for the relationship between volunteer participants and employers (beneficiaries)	Legal basis for the relationship between volunteer participants and employers (beneficiaries) (presentation and video), volunteer service contract template, Volunteer Code of the Kaliningrad Region	Test
	1.4 Motivation to participate in voluntary service and area of activity	Motivation to participate in voluntary service and area of activity (presentation and video)	Test
	1.5 Elements of the mechanism for the practical implementation of voluntary service	Elements of the mechanism for the practical implementation of voluntary service (presentation and video)	Test self-test questions
Summative assessment	Test	Summative assessment rules	Test

- How the volunteer will perform his or her particular task;
- What not to do when performing this task;
- How to handle an emergency or what to do when something unexpected happens;
- What the goals are for the task, and how performance will be evaluated;
- What equipment will be required and how to use it;
- A walk through of the task and coaching while the volunteer tries out the task.

Table 2 Universal competencies (UC) and general professional competencies (GPC)

Competency code	The results of the development of the educational program	Learning outcomes by discipline
UC-3	Ability to socialize and fulfill your role in a team	To know the framework of categories and concepts of the problem of volunteering To be able to apply the acquired knowledge and skills in practical activities To master technologies for organizing volunteer events
GPC-3	The ability to organize joint and individual educational activities of students, including students with special educational needs and disability, in accordance with the requirements of federal state educational standards	To know the requirements of federal state educational standards To be able to organize joint and individual educational and educational activities of students To master technologies for organizing educational and educational activities of students, including students with special educational needs and disability

4 Analysis and Discussion of the Results

In the perspective of enriching the course and attracting students to serve at international scientific and practical conferences, symposiums, congresses with the involvement of foreign participants, we are currently developing a variable part, which will be aimed, among other things, at the formation of intercultural competence of future educators, especially future teachers of English.

It is planned that the course includes six parts. First part of the course is an assessment of intercultural competence of students (A. Fantini test). The second part of the course involves the study of regulatory documents related to the international regulation of volunteering (the Universal Declaration of Volunteering (2001) [16], Article 29 of the Universal Declaration of Human Rights [19]). The third part of the course is dedicated to writing an essay on the topic of "The role of culture in human life" to raise students' awareness of the complexity of culture and develop an understanding of the concept of culture. The fourth part of the course includes an exercise aimed at identifying students' own "cultural backpack" and developing its image. The next part is composed of practical tasks devoted to the study of stereotypes and how to handle them (D.A.E model (Describe, Analyze, Evaluate) by researchers K. A. Nam and J. Condon). The last part includes practical tasks aimed at developing self-regulation in stressful situations, stress resistance, including the situations of interaction with members of different cultures.

Volunteer practices of the Educational and Scientific Cluster "Institute of Education and Humanities" of Immanuel Kant Baltic Federal University.

Table 3 Examples of test tasks

1.1. The history of the volunteer movement in Russia and abroad	1.2. Basic terms and general approaches to volunteering	1.3. Legal basis for the relationship between volunteer participants and employers (beneficiaries)	1.4. Motivation to participate in volunteer service and area of activity	1.5. Elements of the mechanism for the practical implementation of volunteer service
1. In the 18–19 centuries, volunteers were called … • handymen • clergymen • soldiers • rural teachers	1. What is the global sign of volunteering according to the Universal Declaration on Volunteering (2001)? • blue lighthouse of hope • red heart on white background • red V • outlines of palms raised up	1. A volunteer carries out voluntary service: • without salary • without paid vacation • without salary but with paid vacation	1. The manifestation of one's abilities and capabilities, the implementation of human destiny relate to the following motive for volunteering: • Social recognition, sense of social significance • Realization of personal potential • Self-expression and self-determination • Professional orientation	1. What needs to be done first of all to ensure that the public is informed about the opportunities for volunteering? • identify sources of information • determine the target group to which the information will be sent • choose a style of addressing • determine the form of feedback
2. When was the Young Men's Christian Association (YMCA) organized? • 1934 • 1941 • 1844 • 2001	2. Citizens carrying out charitable activities in the form of gratuitous labor in the interests of the beneficiary, including in the interests of a charitable organization, are: • beneficiaries • volunteers • volunteer coordinators • voluntary organizations	2. What is the name of the document that ideologically unites the volunteer movement in a constituent entity of the Russian Federation? Regulations on volunteers • Volunteer Code • Volunteer service contract • Personal book of volunteers	2. Positive reinforcement of one's activities from significant others, a sense of involvement in a generally useful cause relate to the following motive for voluntary service: • Social recognition, sense of social significance • Realization of personal potential • Self-expression and self-determination • Professional orientation	2. The way to determine the optimal solution of economic, managerial and other tasks, by simulating or modeling the economic situation and the rules of behavior of participants is: • training • business game • case method • work briefing

(continued)

Table 3 (continued)

1.1. The history of the volunteer movement in Russia and abroad	1.2. Basic terms and general approaches to volunteering	1.3. Legal basis for the relationship between volunteer participants and employers (beneficiaries)	1.4. Motivation to participate in volunteer service and area of activity	1.5. Elements of the mechanism for the practical implementation of volunteer service
3. When was the International Committee of the Red Cross (ICRC) created? • 1934 • 1863 • 1844 • 2001	3. The form of social service, carried out by the free will of citizens, aimed at the disinterested provision of socially significant services at the local, national or international levels, contributing to personal growth and development of citizens (volunteers) performing this activity, is: • voluntary organization • voluntary service • volunteer coordinators • volunteers	3. In which case a volunteer organization can refuse to cooperate with a volunteer in all or some areas of activity? If the volunteer regularly does not perform the work assigned to him • if a volunteer violates the provisions of the Volunteer Code • if the volunteer does not comply with the work schedule • if the volunteer regularly does not perform the work assigned to him and violates the provisions of the Volunteer Code	3. The opportunity to express oneself, to declare one's life stance, to find one's place in the system of social relations refers to the following motive of voluntary service: • Social recognition, sense of social significance • Realization of personal potential • Self-expression and self-determination • Professional orientation	3. Teaching technique of using a description of real economic, industrial and social situations. Students must analyze the situation, understand the essence of the problems, propose possible solutions and choose the best of them—these is • training • business game • case method • work briefing
4. Who was the founder of the American Red Cross in Washington? • A. Dunant • C. Barton • J. Williams • P. Serezol	4. International Volunteer Day is celebrated: December 15th June 8th June 1st December 5th	4. The duration of the working week for volunteers under 16 years old (if they do not combine work with study): • no more than 24 h • no more than 36 h • no more than 18 h • no more than 10 h	4. Orientation in various types of professional activity, getting a real idea of the intended profession or choosing the area of professional training relate to the following motive for voluntary service: • Social recognition, sense of social significance • Professional orientation • Realization of personal potential • Self-expression and self-determination	4. What training methods can be applied to train a volunteer outside the workplace? (multiple choice) • briefing • lecturing • business games • tasks, arranged in order of increasing difficulty • workplace change • delegation of responsibility method • holding conferences and seminars

(continued)

Table 3 (continued)

1.1. The history of the volunteer movement in Russia and abroad	1.2. Basic terms and general approaches to volunteering	1.3. Legal basis for the relationship between volunteer participants and employers (beneficiaries)	1.4. Motivation to participate in volunteer service and area of activity	1.5. Elements of the mechanism for the practical implementation of volunteer service
5. Which of the rulers of Russia founded the Imperial Philanthropic Society in 1802? • Empress Catherine II • Emperor Alexander I • Emperor Alexander II • Emperor Peter I	5. People receiving charitable donations from philanthropists, the help of volunteers are • beneficiaries • volunteers • volunteer coordinators • voluntary organization	5. Duration of the working week for volunteers from 16 to 18 years (if they combine work with study): • no more than 24 h • no more than 36 h • no more than 18 h • no more than 10 h	5. The acquisition of computer skills, including skills of dealing with various types of equipment, experience of interpersonal interaction relate to the following motive for voluntary service: • Social recognition, sense of social significance • Professional orientation • Acquisition of useful social and practical skills • Self-expression and self-determination	5. What specific activities can be held to find volunteers? • special promotions; • training and other educational activities; • information events • all of the above

Volunteering practices of the Educational and Scientific Cluster "Institute of Education and the Humanities" of Immanuel Kant Baltic Federal University (formerly the Institute of Education of Immanuel Kant Baltic Federal University) began to have a systemic and mass character since 2018, due to the introduction of the discipline "Fundamentals of volunteering and social project planning" in the curriculum for preparing bachelors of pedagogy. Since 2018 more than 300 students have participated in various internal events of the cluster as volunteers and have been awarded letters of gratitude from the head of Educational and Scientific Cluster "Institute of Education and the Humanities" of IKBFU.

The letters of gratitude from the administration of the Educational and Scientific Cluster "Institute of Education and the Humanities" of IKBFU allow students of the cluster to complete the list of achievements and become more competitive in the competition for increased state academic scholarships for special achievements in social activities of IKBFU. Also volunteering enables students to gain experience in socially significant activities, get a record in a volunteer book indicating the number of working hours, find like-minded people and make new acquaintances and connections, enrich their CVs, and receive recognition for their merits. The students of the cluster who have demonstrated their high interest in organization, implementation and realization of social practices and volunteering can be recommended to participate in youth forums, student gatherings, etc.

It should be mentioned that among the volunteer practices online the project "Volunteers of Education" was implemented. Due to the COVID-19 pandemic thousands of regional schoolchildren found themselves in a non-standard educational situation. Educational and Scientific Cluster "Institute of Education and the Humanities" of IKBFU in partnership with the Ministry of Education of the Kaliningrad Region launched an online project "Volunteers of Education". In terms of the project we offered new solutions for the problems that arose during the transition of educational institutions to online learning, such as the need for additional explanation of new material, homework preparation, assistance in developing self-organization and self-regulation skills, organizing leisure and extracurricular activities, etc. [6].

The essence of the project was to improve the methods, forms and content of voluntary service of bachelors of pedagogy for schoolchildren and teachers in the online educational environment. The project is closely related to filling the educational online environment with new content through close social partnership and a new area of volunteering.

5   Conclusions

In conclusion, we believe that the online course "Fundamentals of volunteering" developed by us will be in demand by young people in the future. It not only provides a powerful theoretical basis for understanding the phenomenon of volunteering, but also allows the student to be maximally prepared for the real volunteer practice. In addition, this course is easily scalable, as it has no restrictions on the geography of participants.

We see the prospects for the development of online training for volunteers in the following focus areas:

– Creating a culture that embraces volunteers;
– New approaches to volunteer recruitment;
– Assessing volunteer performance;
– Building and sustaining strong volunteer leadership;
– Managing challenging volunteers and reducing conflict;
– Demonstrating volunteer impact.

References

1. Zyablov, V.A., Gusev, M.A., Chizhikov, V.S.: Clinical features of first-episode psychoses during the COVID-19 pandemic. Consortium Psychiatricum **2**(3), 27–33 (2021)
2. Elsaid, N., El Nagar, H., Kamal, D., Bayoumi, M., Kamel, M., Abuzeid, A., Elewa, S., Hussein, M., Hussein, H., Elshahidy, A., Saleh, J.: Perception of online learning among undergraduate students at suez canal medical school during the COVID-19 pandemic: a cross-sectional study. Egyptian J. Hospital Med. **85**(1), 2870–2878 (2021)
3. Li, J., Wu, C-H.: Determinants of Learners' Self-Directed Learning and Online Learning Attitudes in Online Learning, https://www.researchgate.net/publication/371521360_Determinants_of_Learners%27_Self-Directed_Learning_and_Online_Learning_Attitudes_in_Online_Learning#fullTextFileContent. Accessed 27 June 2023
4. Legacy strategic approach: moving forward. Lausanne: International Olympic Committee, https://library.olympics.com/Default/doc/SYRACUSE/173146/legacy-strategic-approach-moving-forward-international-olympic-committee?_lg=en-GB. Accessed 29 June 2023
5. Suhar'kova, M.P.: Olimpijskie volontery o volonterstve. Vestnik Kemerovskogo gosudarstvennogo universiteta. Seriya: Politicheskie, sociologicheskie i ekonomicheskie nauki **3**, 298–304 (2022)
6. Zelko, A.S., Maslo, V.S.: Organization of work of education volunteers online, https://www.researchgate.net/publication/352657189_Organization_of_work_of_education_volunteers_online. Accessed 25 June 2023
7. Federal Act No. 135 of 11 August 1995 On charitable activities and volunteering, https://www.consultant.ru/document/cons_doc_LAW_7495/. Accessed 27 June 2023
8. Abramova, S.V.: Role of online volunteership in forming social activity of youth. Social'naya rabota: sovremennye problemy i tekhnologii **1**, 17–22 (2020)
9. Podnebesnyh, E.L., Zelko, A.S.: Model of psycho-pedagogical training of students-volunteers in high school. Historical Soc. Educ. Ideas **7**(2), 264–269 (2015)
10. Denisova, O.A., Denisov, A.P., Drobyshev, V.V.: The position of medical students on volunteering: online survey results. Public Health Life Environ. PH&LE **30**(12), 24–29 (2022)
11. Frendo, M.: Exploring the Impact of Online Training Design on Volunteer Motivation and Intention to Act [Michigan State University], https://d.lib.msu.edu/etd/4829. Accessed 25 June 2023
12. Hill, T.G., Langley J.E., Kervin E.K., Pesut, B., Duggleby, W., Warner, G.: An Integrative Review on the Feasibility and Acceptability of Delivering an Online Training and Mentoring

Module to Volunteers Working in Community Organizations, https://www.researchgate.net/publication/355205425_An_Integrative_Review_on_the_Feasibility_and_Acceptability_of_Delivering_an_Online_Training_and_Mentoring_Module_to_Volunteers_Working_in_Community_Organizations. Accessed 27 June 2023

13. In-Person and Virtual Training, https://www.energizeinc.com/training. Accessed 27 June 2023
14. Liu, H., Harrison, Y., Lai, J., Chikoto, G., Jones-Lungo, K.: Online and virtual volunteering. In: The Palgrave Handbook of Volunteering, Civic Participation, and Nonprofit Associations. Palgrave Macmillan, London, pp. 290–310 (2016)
15. Timofeeva, S.V.: The role of student volunteering in the US higher education system. Biznes. Obrazovanie. Pravo **4**(61), 521–525 (2022)
16. Universal Declaration on Volunteering (IAVE, 2001), https://www.iave.org/advocacy/the-universal-declaration-on-volunteering/. Accessed 27 June 2023
17. Concept of volunteering development in Russian Federation up to 2025, https://www.consultant.ru/document/cons_doc_LAW_314804/985421faba1da8d5a7dd327f05ae6cd5f9aa2c4c/. Accessed 27 June 2023
18. Fritz, J.: The Basics of Volunteer Orientation and Training, https://www.liveabout.com/the-basics-of-volunteer-orientation-and-training-2502594. Accessed 27 June 2023
19. Universal Declaration of Human Rights, https://www.un.org/en/about-us/universal-declaration-of-human-rights. Accessed 27 June 2023

Digital Competencies of Student Labor Teams' Members

Galina Zavada⊙, **Maria Reimer**⊙, **Natalia Savotina**⊙, and **Shamil Yakupov**⊙

Abstract The early involvement of young people in professional activities by means of student labor teams opens up opportunities for the formation of competencies that allow them to actively integrate into the world of the digital economy, master modern technologies and be ready to apply them in their professional activities. According to the authors, in the modern scientific community and among practitioners of education, insufficient attention is paid to the opportunities provided by training and work in student teams to form the specialists necessary for the modern economy. The chapter analyzes the possibilities of student labor teams in the formation of digital competencies; a number of features in the requirements of employers regarding digital literacy have been identified. Using the method of theoretical analysis and systematization of scientific sources, a generalization of ideas about the professionally important and personally important qualities of future professionals was carried out, the components of digital competencies in the structure of competencies of fighters of student labor teams were identified, and differences in ideas about the need and possibilities for the formation of various qualities and competencies, including digital ones, in the conditions of SLT were identified. A comparative analysis of the results of surveys of employers and members of student labor teams shows the actual demand for digital skills, both on the part of employers and members of student labor teams, and, according to a number of indicators, the importance of employers is higher than that of future SLT "fighters". This is due to a deeper understanding by employers of labor functions and the specifics of the activity. The results show the presence of a problem field in the field of preparation for work in labor teams in general and in the formation of digital literacy competencies among their members.

Keywords Digital transformation · Student labor teams · Content analysis · Competencies · Nomenclature of qualities

G. Zavada (✉) · S. Yakupov
Kazan State Power Engineering University, Krasnoselskaya str., 51, Kazan, Russian Federation
e-mail: g.zavada@mail.ru

M. Reimer · N. Savotina
Kaluga State University, 22/48 Stepan Razin Street, Kaluga, Russian Federation

1 Introduction

Education at the university is aimed mostly at acquiring theoretical knowledge, skills and abilities by students. The modern realities of the economy are developing in such a way that "flexible" (personal) skills are highly valued in production, so the issue of acquiring by students not only professional (PVC), but personally important qualities (PSC) while studying at a university is becoming more and more significant in the educational system. In addition, the need to acquire digital competencies is becoming increasingly acute. Modern society is focused on information technology in almost all spheres of life. According to researchers, the trend of informatization, digitalization, is the most transforming modern system of education and labor. In studies of the Organization for Economic Cooperation and Development [1], the dominant position of digital competencies among the competencies of the twenty-first century is noted. Digital transformation covers various professions, requiring new competencies from employees, the ability to work in a changed environment. This is confirmed by a number of data, so, according to [2], changes associated with digitalization in 2019 were noted by 47% of workers in various professions. The ongoing transformation of professional activity entails the restructuring of programs and technologies for training and preparation for it, as well as the design, in particular, of updated competence models of university graduates that would provide relevant education received at a university.

Many employers develop a certain portrait of a student as an ideal model that has different qualities; on its basis, a requirement is formed that upon graduation from the university, the student should already be endowed with the necessary qualities, depending on the direction of the company and the position held. Various studies speak for this. The authoritative magazine for entrepreneurs "Inc. Russia" conducted a study among employers. According to the results of the survey, it turned out that more than half of the companies −71%—hire young specialists based on the assessment of their personal competencies and how they correspond to the corporate culture and values. As for individual personal or professional qualities, the majority of employers—more than 83%—named the ability to work in a team and focus on results as the most important qualities, about 59% of respondents named adaptability among the most important qualities [3]. After comparing more than 30 similar studies, we note the similarity of the "set" of the necessary qualities and competencies that are required from a university graduate. Among them, the need for the formation of non-specialized, but important for the implementation and building of a career skills, flexible, independent of the specifics of the profession, is actualized. These so-called soft skills, in the ideal embodiment, provide high labor productivity, possibly less "energy consumption" of professional activity. The skills of a university graduate, which incorporate the key trend of modern education, include skills of working in the information environment, consisting of the ability to work on the Internet, carry out information search, implement communication (including professional) using digital means, etc. Surveys of employers record a gap not only in terms of basic digital competencies (the ability to search for information, digital communication skills), but also in specific professional skills [4]. An analysis of the opinions of employers and various studies convinces of the upcoming digital transformation of the requirements for participants in labor processes, and, consequently, the transformation of the training process [5, 6].

Education in higher education creates a lot of opportunities for the formation of the necessary qualities and skills, for personal and professional development both during the educational process, through the alternation of various disciplines, teaching technologies, and in extracurricular activities: through public organizations, student governments, trade union committees, volunteer organizations, creative and sports clubs, etc. A whole body of research analyzes various educational technologies, the use of which develops digital skills [7, 8]. The greatest interest in terms of the ongoing research is caused by student associations focused on the formation of primary labor skills.

One of the largest youth organizations in Russia and a number of post-Soviet countries are student labor teams (SLT). This is an organization whose purpose is the employment of students during the vacation period in various areas of activity (pedagogical, construction, energy, service, agricultural, medical, etc.). The organization first announced itself in 1959, in Moscow, Moscow State University, when students of the Faculty of Physics went to a student construction site in Kazakhstan. Between 1965 and 1991, about 13 million students took part in the SLT. It can be noted that the functioning of student labor detachments was very important in a planned economy, the state also switched to the planned formation of Student detachments. Since 2003, in the Russian Federation there has been an organization "Russian student teams", which is the largest youth public organization. More than 240,000 students annually go through the school of student brigades in the country. In the Republic of Tatarstan, for the summer of 2023, there are 4,985 student labor teams (including pedagogical, service, construction, medical, industrial, agricultural, guide teams), in which 11,985 students take part; the movement is being revived in other regions of the country, in particular in the Kaluga region—construction, medical, pedagogical and agricultural teams, a total of 220 people, which gives reason to talk about the advisability of considering the possibilities of SLT for the formation of the required competencies [9, 10].

According to the authors, in the modern scientific community and among practitioners of education, insufficient attention is paid to the opportunities provided by training and work in student teams to form the specialists necessary for the modern economy. In general, the analysis of scientific literature shows that the quality of professional activity is a derivative of the personal qualities of the subject of activity; research recognizes the value of digital transformation of competencies; it necessitates changes at all levels and forms of obtaining professional skills, which we can include training for activities in service stations related to digitalization; and also the problem of corresponding changes in the nomenclature of professionally important qualities of the participants of student labor teams is formulated.

2   Materials and Research Methods

To solve the problem posed, we used the following research methods: theoretical analysis of the problem based on the study of pedagogical, methodological, psychological and special sources; comparative analysis of advanced scientific and pedagogical experience related to the functioning of the SLT and the development of personally and professionally important qualities in its conditions; content analysis; survey of employers and service station participants.

Using the method of theoretical analysis and systematization of scientific sources, a generalization of ideas about digital competencies in the structure of competencies of fighters of student labor teams was carried out. Comparative analysis made it possible to study the opinions and experience of domestic and foreign researchers on the problem. A survey of participants in Student Labor Teams, their leaders and employers (more than 170 participants) revealed differences in perceptions about the need and possibilities for the formation of various qualities and competencies, including digital ones, in the conditions of SLT.

3   Results

A significant number of the considered studies are devoted to the formation of individual qualities or groups of qualities associated with the effectiveness of the implementation of professional activities. Prikhodchenko E. I. and Kapatsina N. N. note that the main sources of sustainable global professional success in modern conditions are creativity and innovation. The formation of such qualities requires a specific approach to the organization of the educational approach, in particular, the authors propose a phased introduction of the lateral marketing technique in the training of future engineers [11].

In [12], the emphasis is placed on the quality of "personal ambition" as a professionally important and as a factor in the success of the activity. Researchers Lapshova A. V., Urakova E. A., Mikhailenko D. M. note that the process of forming professionally important qualities can be carried out in various ways, through a specially designated academic discipline; in an end-to-end way, through all academic disciplines; through the system of extracurricular activities and additional education; through the creation of a special practice-oriented student space [13].

As noted by Zeer E. F. and Zavodchikov D. P. the most important quality of a modern specialist is "dynamic professionalism", which can be represented as a reflection of his "universality" [14]. In this regard, the importance of "soft" competencies, skills—communication and teamwork, flexibility, readiness for learning, etc. is emphasized. This approach is aimed at the constant self-realization of the employee, significantly expanding his capabilities and professional areas; Available studies in developed countries show that the salary is higher for those who are good at precisely such "soft" skills [15].

An interesting fact is that employers highly appreciate the importance of such skills and their corresponding qualities. The chapter [16] concluded that critical thinking skills, creative thinking, negotiation skills, the importance of emotional intelligence and cognitive flexibility are in demand. This opinion is reflected in the study [17], and also correlates with the results of our survey of employers.

The studies of many authors note the already undeniable transformation of labor functions, the cause of which is their digitalization. M. Ford notes: "computers are better and better able to perform highly specialized, routine and predictable tasks and, it is likely that they will soon surpass many people who are now engaged in such work in this skill" [18]. Note that in the context of this chapter, we use the concepts of "digital literacy" and "basic digital skills" synonymously, since the totality of basic digital skills actually forms the digital literacy of a modern person. Digital competence is a combination of several digital skills and acquired knowledge for continuous application in professional activities. At the same time, as Laara E., Deursena A. J., and others point out, digital competence covers information management, collaboration, communication and sharing, content and knowledge creation, ethics and responsibility, assessment and problem solving, and technical operations" [19]. An interesting review by Pettersson F., which notes that digital competence in the educational context has been considered in international studies over the past few years in terms of society development, organizational infrastructure, strategic leadership, as well as the skills of teachers and their teaching technologies. The author also emphasizes that digital competence can be considered as an organizational task, which is influenced by several factors built into the broader organization of the educational process [20].

Considering labor teams in the context of the formation of digital competence of students in preparation for work in them, we note that they provide a significant opportunity for students of educational institutions to acquire the first professionally important qualities and work experience. Members of student groups learn to work in a team with their peers, gain experience in communicating with the employer, show their leadership skills, and receive an additional working profession before employment [20]. Fighters of student labor teams automatically become more competitive than their peers due to acquired professional and personal qualities.

Researchers and practitioners emphasize a number of tasks, the implementation of which is carried out by the SLT: temporary employment and improving the material well-being of students; labor education of students; formation of a personnel reserve of the economy, etc. [21, 22].

The experience gained in SLS makes students more "expensive" in the labor market. At the same time, the specifics of the student's work in SLS, associated primarily with their temporary nature, raises the question of the adequacy of the development of the considered qualities, their specificity and set. Despite a significant number of works on this topic, the theoretical analysis of research and the study of the experience of assessing professionally important and personal qualities made it possible to highlight a certain contradiction between the need of society and the labor market for competitive specialists with the necessary level of digital qualities and skills, and the specifics and influence on their formation of student labor teams.

Most often, researchers talk about the formation of PVK as a concomitant result of employment in a student labor teams, the qualities themselves, their nomenclature and specificity did not become a separate subject of research. In this regard, the problem of identifying the nomenclature of professionally important qualities of participants in student labor teams is being updated, as well as a description of the features of these qualities associated with the specifics of the service stations themselves, and the identification of digital content in the nomenclature.

There are various interpretations of the category "professionally important qualities" (PVK), for example, as "individual qualities of the subject of activity that affect the effectiveness of the activity and the success of its development" (according to V. D. Shadrikov) [23]. This approach is interesting in that it emphasizes the importance of the ITC both for the productivity, efficiency, and effectiveness of the professional activity itself, and for the quality of the preparation process for it. Undoubtedly, the fact that PVC is closely associated with personal qualities (LC) gives us another important issue—the relationship and interdependence of professionally important and personal qualities, including a university graduate.

An analysis of scientific research has shown a variety of approaches and formulations, both in understanding PVK and LC, and in their content. In this case, it seems relevant to conduct a theoretical analysis of the existing material in order to find the desired formulations. We used content analysis, which, based on flexibility and the possibility of formalization, is used as an effective method of practical analysis of the text by counting the concepts being analyzed and determining the connecting components between them in a qualitative and quantitative ratio. The study was carried out in five stages: the first—the search and selection of literature on personal qualities; the second—a selection of the desired formulations and their systematization; the third is the content analysis of the selected texts; fourth—analysis-synthesis of the results of content analysis to obtain intermediate formulations; the fifth is incomplete induction, when a general conclusion is made based on consideration of several intermediate formulations [24]. As the desired formulations of personal qualities, the author's definitions of A. A. Verbitsky, V. D. Shadrikova, A. V. Ponomareva, V. I. Turanina and others [6, 14, 23, 25]. As a result of the work carried out, the following data were obtained (the enumeration indicates the concepts and the frequency of occurrence of this concept in the definitions): social—6; biological—2; structural—4; components—3; personality (ny) —8; stable—5; states—4; mental—3; processes—5; properties—5; behavior—8; natural environment—4; quality—2; set—2; devil—2; features—4; set—2; people—7; set—2; character—3; sphere—2; expressing—2; which—5. Based on these data, we can formulate the following definition: personal qualities are a set of socio-biological structural components of a personality, which is expressed by stable states, mental processes, behavior, properties, character traits, human characteristics in the natural environment.

In addition, it can be noted that the concept of "personal qualities of a graduate of higher education" is multifaceted. The study of the content of "personal qualities of a graduate of a higher school allows us to identify their main groups:

- qualities of sociability (communication, ability to work in a team, respect for the personality of another person);
- leadership qualities (ability to take risks, desire for continuous self-development, desire for professional growth, courage, determination, persuasiveness);
- intellectual qualities (good memory, sufficient erudition);
- qualities of creativity (creativity, initiative, ability to think critically);
- qualities of industriousness (responsibility, purposefulness, discipline, organization);
- psychophysiological qualities (stress resistance, adaptability);
- moral qualities (decency, honesty, spiritual freedom, humanism, self-esteem, desire for self-development).

Conducting the study of the concept of "professionally important qualities" according to the presented algorithm, we relied on the research of V. D. Shadrikova, A. A. Derkach, E. F. Zeera, V. A. Tolochka and others [11, 22, 24]. Among the most frequently used terms in the definitions were identified (we indicate as the number of repetitions of key terms decreases): "professional", "activity", "qualities (properties)", "success", "necessary", "individual", "efficiency/performance", "quality (activities)", "process", "performance", "professions", "labor". The processing of the identified terms makes it possible to define the concept of personal PVK as follows: professionally important qualities are the individual qualities of a person necessary for the effective, efficient and high-quality performance of professional activities. The content of the PVK content varies depending on the types of professional activity, based on specific labor functions.

In general, in the pedagogical literature, three main approaches to the formation of the content of the SVC can be identified, which are presented in Table 1 (developed by the authors on the basis of [25, 26]).

Summarizing the data given in Table 1, we note that all the authors emphasize the dependence of the content of the PVK on labor functions and tasks; in a detailed description of the content, they are focused on variability based on the profession,

Table 1 Approaches to the description of PIQ content

Approach	Essence	Authors
PIQ as a quality system	Provides a high probability of successful professional development and activities of a specialist (synonymous with "competence")	Shadrikov V. D., Mitina L. M. and etc.
PIQ as a psychological concept	Individual dynamic characteristics of the personality: mental, physiological and psychomotor properties, abilities that meet the requirements for the personality of a specialist in a particular profession and contribute to its successful mastery	Platonov K. K., Markova A. K. and etc.
PIQ as individual qualities	Along with knowledge, skills and abilities, are an integral part of the competence and professionalism of a specialist	Petruk G. V., Zavalishina D. N. and etc. др

but there is no single approach to describing the structure. You can find a description of "qualification characteristics" (V. A. Slastenin); enumeration of the actual qualities that, according to the developers, affect the efficiency and productivity of labor; describe PVK as derived from knowledge, skills and abilities in the professional field or from professional functions. In our opinion, reliance on professional functions is the most objective. A well-written professional standard with a description of generalized and specific labor functions gives the most complete understanding of the planned learning outcome and, as a result, the necessary content and learning technologies. Based on the complex of functions, it is possible to compose a system of professionally important qualities, the presence of which allows a specialist to effectively perform activities. In various studies, one can see stable groups of professionally important qualities for a certain group of specialties or activities.

In a scientific study, Moslan T. S. [27] presents a list of professionally important qualities of teachers, which we can use when designing training programs for participants in pedagogical labor teams. The list includes 113 characteristics, which are presented as proper personal qualities, for example, a sense of humor and diction; skills: use innovative technologies in teaching, etc.; psychological characteristics: long-term memory, cognitive flexibility, etc. Among the considered STCs, 13 characteristics related to digital competence can be distinguished:

Possession of ICT competencies in the field of creating and administering websites.
Possession of ICT competencies in the field of creating and administering websites.
Knowledge of information security skills.
Knowledge of the basics of web design.
Knowledge of modern video communication tools (skype, zoom).
Possession of modern means of on-line documentation (google, bitrix, office.live, document.on-line).
Possession of modern forms and methods of teaching (on-line training).
Proficiency in viewing, listening and creating video and audio files.
Possession of development tools for on-line testing, voting and polls.
Ability to create and edit video lectures.
Information susceptibility.
Ability to use innovative technologies in teaching.
Ability to use modern forms of knowledge control (on-line testing).
Ability to work with digital educational platforms.

Researchers Krashakova T. Yu. and Tuber I. I. [28], actualizing the problem of training future construction technicians for the digital economy, are based on the key competencies proposed by the federal project "Personnel for the Digital Economy". In particular, they include:

Communication and cooperation in the digital environment.
Information and data management, etc.

Noting the over-professional nature of these qualities, the authors emphasize the cross-cutting nature of the formation, which allows us to talk about the possibilities of labor teams in their development and acquisition.

Based on the theoretical and methodological analysis of the literature and the professional sphere, we developed a map for an express survey of members of student labor teams, as well as employers, and based on it, a nomenclature of professionally important qualities that can be successfully formed in the conditions of student teams was determined. The purpose of this technique was to identify LS and PVK, which are guaranteed to be formed in the conditions of SRT, as well as to determine the specific features of the system of identified qualities. The survey involved 44 employers and 127 "fighters" of student groups in Kazan, the Republic of Tatarstan, Kaluga and the Kaluga region. Respondents were offered a form with 30 PTCs, the list of which was compiled on the basis of a summary of studies on the nomenclature of professionally important qualities of various areas of professional activity, and among them the most common and suitable for all types of professions were highlighted. Respondents were asked to evaluate on a ten-point scale how important the presented characteristics are in the upcoming work activity (indicator 1), based on the profile of the student group) and to what extent these STCs are already being formed in the proposed training programs, in preparation for work in the service station (indicator 2). Among the selected thirty characteristics, six were associated with digital skills: computer and office equipment skills (C1), ability to work with information (C2), ability to work with various digital resources (C3), possession of ICT competencies in the field of creating and administering websites (C4), knowledge of information security skills (C5), the ability to communicate using various digital tools (C6). In the context of the proposed work, we will focus only on these six characteristics and analyze the correspondence between significance and formability, according to employers and team members themselves (see Figs. 1 and 2). The graphs show the arithmetic mean estimates of the selected skills for two indicators—the significance in the professional activity corresponding to the work of the labor team (1) and the formation during the process of preparing for work (2).

A comparative analysis of the results shows the real demand for digital skills, both on the part of employers and members of student labor teams. First of all, we note that for all indicators, the assessments that characterize how digital skills are formed in the process of preparing for activity are lower than those that indicate the significance of the skill in the profession. In our opinion, this information shows the insufficiency of the content and technologies of the training program for the formation of these skills; in fact, these skills are not formed in the training programs.

According to four indicators: computer and office equipment proficiency, ability to work with information, ability to work with various digital resources and possession of ICT competencies in the field of creating and administering websites, assessments of significance in future professional activity are higher for employers, for example, the importance of "the ability to work with information" is estimated by employers at 9.15 out of 10, and students at 7.94 points. Of all the skills identified in this work, only "mastery of information security skills" and "ability to communicate using various digital means" are rated higher by students than employers (respectively 9.4 versus

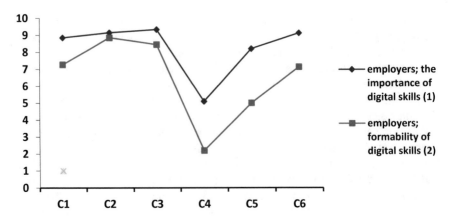

Fig. 1 Differences in the importance of digital skills according to employers

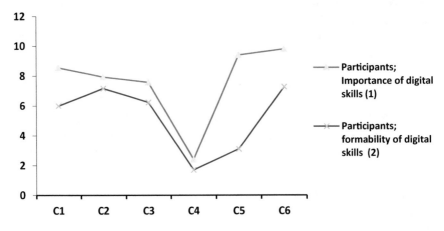

Fig. 2 Differences in the importance of digital skills according to SLT participants

8.2 and 9.8 versus 9.12). A similar situation develops when analyzing opinions on those formed in the process of preparing for work in student labor teams. We can observe the excess of the importance of five digital skills in the opinion of employers over the opinion of students: computer and office equipment skills (7.27 vs. 6), the ability to work with information (8.85 vs. 7.17), the ability to work with various digital resources (8.45 vs. 6.24), the possession of ICT competencies in the field of creating and administering websites (2.2 vs. 1.7), and the knowledge of information security skills (5 vs. 3.1). This is due, in our opinion, to a deeper understanding by employers of labor functions and the specifics of activities. It is also necessary to pay attention to the low significance of the skill "possession of ICT competencies in the field of creating and administering websites" in both samples, the reason for which may be the absence of this labor function in future work in the student team.

In general, the results show the presence of a problem field in the field of preparation for work in labor teams. Based on the ranking by the importance of professional qualities, both on the part of employers and participants, it is possible to more reasonably form didactic units of the content of the training process and select appropriate learning technologies, depending on the required competencies. Feedback on the extent to which these skills are already being formed in the proposed training programs, when preparing for work in the service station (indicator 2), is a guideline for the organizers of the training programs for their timely updating and adjustment.

4 Discussion

The rapid development of new technologies and the expanding digital transformation of skills in demand on the labor market have predetermined the need to study specific content, both in the nomenclature of professionally important qualities of future specialists, and in their training programs. In the theoretical block of the work, more than 50 sources were analyzed in terms of the structure and content of the concepts of professionally important qualities, the content analysis made it possible to clarify a number of pedagogical terms. Comparison of various author's developments of PVK nomenclatures in a number of areas of training shows the intersection of many "digital skills".

Since the revival of the activity of student labor detachments, interest in studying their role in the development of a young person has increased significantly. In the scientific citation database of the Russian Federation for the period since 2003 (the date of the beginning of the functioning of the SLT it is possible to analyze about twenty thousand publications of various levels, considering the current aspects of the functioning, development and history of this youth phenomenon. Having examined in more detail 378 works since 2015, we note that most of the studies emphasize the educational potential of activities in student labor groups. Researchers analyze and prove the serious impact of work in the SRT on the self-esteem of the individual, his self-development and value orientations; the socio-cultural potential of labor groups is emphasized, the formation of various moral qualities: citizenship and patriotism, leadership qualities, healthy lifestyle skills, etc. The second most frequently mentioned position in scientific research is the role of SLT in the professional development of young people. Motivational aspects of labor activity in detachments, the specifics of professional adaptation and professional self-determination are studied; specificity of employment of student youth. However, as the analysis of the works shows, there are practically no studies that reflect the specifics of the "digital economy" and its impact on professional activities in labor teams. Only in the last two years have works appeared in the proposed context, where both the functioning of detachments and the specifics of training are studied. From a practical point of view, the work of Skvortsov D. E. is interesting, offering an automated information system for business processes of a linear student team (report generation, database management, selection of educational resources, etc.) [29]. The digitalization of the management system of the functioning process requires a change in approaches to the management of service stations, specialized training of teachers and commanders of student labor teams.

The issues of forming soft-skills in general, "skills of the future", are becoming relevant. The researchers emphasize that, according to a number of indicators, members of student groups are in a more advantageous position compared to other university students. Kazantseva A. V. and Syryamkina E. G. prove that such skills that give an advantage include the skills of self-organization, self-development, development and implementation of projects, as well as reflection skills [30]. Attempts are being made to systematize the actual "skills of the future" of a modern specialist in the labor market and to determine the role of student labor teams in their formation. Ponomarev A. V. and Kostin N. A. note that modern researchers pay little attention to the role of SRT in the formation of such skills, they offer their system based on the data of the Agency for Strategic Initiatives; distinguish the following skills of the future: critical and creative thinking, strategic thinking, systems thinking, communication, cooperation, self-organization, self-regulation; offer an approach to designing a program for the headquarters of student teams to form the skills of the future [31].

The development of a digital society and digital culture requires increased attention at all levels and areas of education to the problem of young people's readiness for life and profession in a changing environment. The urgency of the problem is emphasized by Simonova S. A. [33], which notes that Russian student groups have a significant potential in creating values as an effective way to form the information security of the student environment and a critical attitude towards information manipulation.

In general, it can be noted that researchers, describing the importance of labor teams for professional education, adaptation and formation of primary labor skills, practically do not take into account modern trends in the digital economy. The skills of working with digital content and the skills of digital communication are outside the scope of research, are not reflected in the description of the content of training in general, the possession of digital skills in working with information.

5 Conclusion

The results of the survey showed that digital skills are in demand both on the part of employers and on the part of participants in the movement of labor groups, therefore, it is necessary to correlate training programs with the tasks set. Analysis and generalization of experience led to the conclusion that the potential of student labor teams in the formation of the declared skills and abilities was not realized. At the same time, our work does not give a complete picture of the demand for digital skills by representatives of labor groups of different directions. It is necessary to study invariant and variable digital competencies for the specifics of the profession. So far, invariant elements of training programs for future members of labor teams have not been developed, the implementation of which will contribute to the development and formation of digital skills, will quickly provide the labor market with specialists with the required competencies, and will contribute to a full-fledged digital transformation of the economy.

References

1. OECD: Skills Matter: Further Results from the Survey of Adult Skills, OECD Skills Studies, OECD Publishing, Paris (2016). https://doi.org/10.1787/9789264258051-en
2. Brolpito, A.: Digital skills and competence, digital and online learning. European Education Foundation. [electronic resource]. URL: https://www.etf.europa.eu/sites/default/files/2019-08/dsc_and_dol_ru_0.pdf. Accessed 13 June 2023
3. Research: personality and education are the most important qualities in a young specialist for an employer. [electronic resource]—URL: https://incrussia.ru/news/dream-young-career/. Accessed 13 June 2023
4. Sicilia, M.-A., GarcíaBarriocanal, E., Sánchez-Alonso, S. et al.: Digital skills training in higher education: insights about the perceptions of different stakeholders. 6th International Conference on Technological Ecosystems for Enhancing Multiculturality (TEEM). New York: Association for Computing Machinery, pp. 781–787 (2018). https://doi.org/10.1145/3284179.3284312
5. Gusev, A.A.: Digitalization of labor relations and its impact on labor productivity and the cost of companies. Taxes. Right. No. 4. S. 18–23. (2019). [electronic resource]. Access mode: https://cyberleninka.ru/article/n/tsifrovizatsiya-trudovyh-otnosheniy-i-ee-vliyanie-naproizvoditelnost-truda-i-stoimost-kompaniy. Accessed 10 June 2023
6. Ponomareva, O.Ya.: The choice of technologies for the development of soft skills by specialists in the context of digitalization [electronic resource]. Ponomareva, O.Ya., Gorkun, M.N., Kozlov, A.S. Access mode: http://inper.academy/wpcontent/uploads/2019/05/DSEME-2018_Conference-Proceedings.pdf, https://elibrary.ru/item.asp?id=39275306. Accessed 10 June 2023
7. Armah, J., Westhuizen, D. embedding digital capability into the higher education curriculum: the case of Ghana. Univ. J. Educ. Res. **8**(2), 346–354 (2020). https://doi.org/10.13189/ujer.2020.080203
8. Alt, D., Raichel, N.: Enhancing perceived digital literacy skills and creative self-concept through gamified learning environments: insights from a longitudinal study. Int. J. Educ. Res. Nr 101. Article number 101561 (2020). https://doi.org/10.1016/j.ijer.2020.101561
9. Russian Student Squads. [electronic resource]—URL: https://trudkrut.rf/o_rossiyskikh_studencheskikh_otryadakh.htm. Accessed 10 June 2023
10. Student groups are being revived in the Kaluga region. [electronic resource]—URL: https://kgvinfo.ru/novosti/obshchestvo/v-kaluzhskoy-oblasti-vozrozhdayut-studencheskie-otryady/.Accessed 10 June 2023
11. Prikhodchenko, E.I., Kapatsina, N.N.: Generating new ideas as a professionally important quality of engineering educators. Civil Defense Acad. J. **3**(23), 127–131 (2020)
12. Panfilova, A.V.: Personal ambition as a professionally important quality of a modern specialist. Int. J. Exp. Educ. 6–1, 15–16 (2014)
13. Lapshova, A.V., Urakova, E.A., Mikhailenko, D.M.: Mechanisms of formation of professionally important qualities of university students. Problems of Modern Pedagogical Education. No. 72–3. [electronic resource] (2021). https://cyberleninka.ru/article/n/mehanizmy-formirovaniya-professionalno-vazhnyh-kachestv-studentov-vuza. Accessed 5 June 2023
14. Zeer, E.F., Zavodchikov, D.P.: Identification of universal competencies of graduates by the employer. Higher Educ. Russia **11**, 39–45 (2007)
15. Ng'ambi, D., Bozalek, V.: Editorial: Massive open online courses (MOOCs): Disrupting teaching and learning practices in higher education. BJET British J. Educ. Technologi **46**(3), 451–454 (2015), [Electron. resource]. Access mode: https://doi.org/10.1111/bjet.12281. Accessed 10 June 2023

16. Kostin, N.A.: Portrait of a soldier of a student construction team in the context of the skills of the future. Kostin, N.A., Ponomarev, A.V.. Text: electronic. Innovative potential of youth: citizenship, professionalism, creativity: collection of scientific chapters of the International Youth Research Conference (Yekaterinburg, November 24, 2020). Yekaterinburg: Publishing House Ural. un-ta, S. 257–261. (2020). [e-resource]—URL:: http://hdl.handle.net/10995/97481. Accessed 15 June 2023

17. Sorokopud, Yu.V., Amchislavskaya, E.Yu., Yaroslavtseva, A.V.: Soft Skills ("soft skills") and their role in the training of modern specialists. World Sci. Culture Educ. 1(86). 194–196 (2021)

18. Ford, M.: Robots are coming: technology development and a future without robots. Ford Alpina "Digital", 261 p. (201)

19. van Laara, E., van Deursena, A.J., van Dijk, J.A., de Haan, J.: The relationship between 21st-century skills and digital skills: a systematic literature review. Comput. Hum. Behav. **72**, 577–588 (2017). https://doi.org/10.1016/j.chb.2017.03.010

20. Pettersson, F.: On the issues of digital competence in educational contexts—a review of literature. Educ. Inf. Technol. **23**, 1005–1021 (2018). https://doi.org/10.1007/s10639-017-9649-3

21. Khovrin A.Yu.: Student brigades as a subject of the implementation of the state youth policy: sociological and managerial analysis: dis. ... candidate of sociological sciences: 22.00.08. Moscow, 205 p. (2003)

22. Gnatyuk, M.A., Krotov, D.V., Samygin, S.I.: Russian student teams as a resource for socio-professional self-determination. Gum. Sots-Econ. Soc. Sci, 31–34 (2017)

23. Shadrikov, V.D.: Problems of systemogenesis of professional activity. M. : Logos, 192 p. p. 86 (2007)

24. Matushansky, G.U., Kudakov, O.R., Zavada, G.V.: Structure and content of the competency-based approach to the training of a professional specialist. Bulletin of the Kazan Technological University. No. 6 (2009). https://cyberleninka.ru/article/n/struktura-i-soderzhanie-kompetent nostnogo-podhoda-k-podgotovke-professionala-spetsialista. Accessed 14 June 2023

25. Tolochek, V.A.: Competence-based approach and PVK-approach: opportunities and limitations. Bulletin of St. Petersburg State University. Series 16: Psychology. Pedagogy. 2019. №2. URL: https://cyberleninka.ru/article/n/kompetentnostnyy-podhod-i-pvk-podhod-vozmoz hnosti-i-ogranicheniya. Accessed 10 June 2023

26. Mospan, T.S.: Individual educational routes as a means of forming professionally important qualities of students of a pedagogical university. Mospan, T.S., Timoshenko, A.I., Ivanova, E.N. Professional Educ. Russia Abroad **3**(35), 80–85 (2019)

27. Mospan, T.S.: Formation of professionally important qualities of future teachers to work in a digital educational environment. https://kemsu.ru/upload/iblock/53f/53f94f1b5860f53e2f373 0de33188d59.pdf. Accessed 10 June 2023

28. Krashakova, T.Yu., Tuber, I.I.: Ways of forming the key competencies of the digital economy among future civil engineers. Innovative development of vocational education. No. 3 (31) (2021). https://cyberleninka.ru/article/n/sposoby-formirovaniya-klyuchevyh-kompetentsiy-tsi frovoy-ekonomiki-u-buduschih-tehnikov-stroiteley. Accessed 24 June 2023

29. Skvortsov, D.E.: Automated information system of a linear student team. Certificate of registration of the computer program 2023616337, 03/24/2023. Application No. 2023614693 dated March 12, 2023

30. Kazantseva, A.V., Syryamkina, E.G.: Formation of soft skills among students and members of Russian student groups: a comparative analysis. Organization of work with youth. №3 2022. URL: elibrary_50397366_52520882.pdf

31. Ponomarev, A.V., Kostin, N.A.: Formation of the skills of the future among student groups. PRIMO ASPECTU. No. **3**(47), 55–59 (2021)

Energy Education in the Era of Digital and Socio-Economic Transformations: Kazan State Power Engineering University Experience in the Formation of Competencies

Radmila Khizbullina

Abstract Currently, in various scientific and methodological discussions there are polar opinions about the development of educational systems, forms and methods. The chapter considers the successful experience of Kazan State Power Engineering University in introducing advanced forms and methods of training of young specialists for the energy industry of the Republic of Tatarstan in modern socio-economic conditions. The chapter provides the result of the implemented experience of Kazan State Power Engineering University on the formation of digital competencies (digital general professional and universal competencies) in the practice of training specialists in the field of training "Economic Sociology and Marketing." In order to analyze the peculiarities of the use of digital educational technologies in the educational process of higher education, an author's sociological study was carried out on the basis of the Kazan state power engineering university. As a method, a survey of students was chosen, which involves block structuring of the toolkit in order to determine some indicators that mediate the influence of digital and information and educational technologies in the educational process among students. Results of applied sociological research conducted in Tatarstan in 2016–2019 within the framework of the study of socio-economic activity, peculiarities of formation of life strategy trajectories of talented youth of Tatarstan made it possible to determine socio-political, socio-economic and socio-cultural metrics. Thus, the chapter examines and attempts to solve the problem of digital transformation of education in the field of energy by defining and practically introducing groups of digital competencies among future specialists.

Keywords Digital transformation · Energy education · Digital competencies · Young specialist · General professional competencies · Universal competencies · Data analysis

R. Khizbullina (✉)
Kazan State Power Engineering University, Krasnoselskaya St. 51, 420066 Kazan, Russian Federation
e-mail: mine_post@inbox.ru

1 Introduction

1.1 Relevance

Today, in the scientific literature, consideration of the socio-economic transformations and "new" economy features is often associated with the study not only of the perspective of the transformation of the institute of economics, but also with the question—"are changes affecting all spheres of public life so fundamental?" [1].

The result of "socio-economic transformations" era appearance seems to be debatable. In the author's opinion the information and industrial structure is a new and leading quality of the socio-economic development. It determines the parameters of the dominant global infrastructure. At the same time, production can still be considered as the core of the new infrastructure. Its main profits are derived from the countries where capital is rushed, where highly skilled personnel are concentrated. Also technically equipped scientific centers are created there [1].

In this regard, it seems interesting to understand the development of new economy as a result and stimulator of the development of competitiveness at the social level.

Transformational processes taking place in modern society inevitably lead to the changes in socio-economic relations in professional education. Structural and institutional changes of the education system is among the most important tasks to achieve high quality training of a competitive specialist. For example, it is true that the development of competitiveness, as the result of vocational education, forms human capital in the modern economical conditions. On the other hand it acts as the dominant factor in the development of both the individual and the economic system of society as a whole.

The conditions, factors and motives of youth orientation from the point of view of their competitiveness, support of social enterprises in the regions of Russia are considered in the works of modern Russian [2, 3] and foreign authors [4]. The conclusion proposed by V. Yu. Melikhov when researching the formation of a post-industrial economy in Russia about the need… "in relation to modern the (intellectual) stage of the formation of a post-industrial society to supply the structure of human capital with a creative and intellectual component" [5] seems to be very interesting. This will allow the subject of economic activity not only to reproduce and accumulate knowledge, but to achieve their multiplicative increase. It means the effect of creating "new" knowledge in the process of its repeatedly increasing volume and quality is both in the process of training and in subsequent implementation. The constant multiplicative increase in knowledge is defined as a necessary condition for the transition to the innovative and intellectual stage of a post-industrial society.

Today, the economy of Tatarstan, as well as at last decades, puts forward and try to solve the problems of improving the quality of life of the population. It means developing the social sphere and services on the basis of the Tatarstan Socio-Economic Development Program till 2030 [6]. Since the development of social facilities and projects is included in this Program, the dissemination of the experience of an effective higher school system is updated in terms of training and developing the competencies of young specialists in modern conditions.

Currently, one of the priorities for the development of the global production industry is the creation and use of smart energy systems (smart grids) that combine achievements in the field of energy and information technologies. The energy sector also needs specialists who can plan, develop and apply innovative solutions in different settings. Here it is worth referring to the experience of Kazan State Power Engineering University, which is one of three specialized energy universities in our country (the other two are Moscow Energy Institute (Technical University) and Ivanovo State Energy University). It occupies one of the leading places in the region in terms of education, technical equipment and conditions for scientific work and the educational process. The CESG is training specialists in 14 areas of bachelor's and master's degrees and in 11 areas of training of certified specialists (31 specialities) in day, correspondence and correspondence forms of education.

Kazan State Power Engineering University trains specialists for power systems of the Volga region, as well as for countries near and far abroad. Students are practiced at energy enterprises in Kazan, the Republic of Tatarstan and Volga region. Today about nine thousand students and graduate students from various regions of the Russian Federation, CIS countries, Asia and Africa study at KSPEU. On the basis of KSPEU, a research institute of energy problems was founded in order to solve problems in the fields of heat and electricity, electrical engineering and electronics, environmental protection and rational use of resources of the Republic of Tatarstan, Volga region and the Western Urals. In terms of the volume and level of scientific work performed, together with students, undergraduates, graduate students, young scientists, KSPEU is one of the best universities in the Russian Federation. The University is equipped with a state-of-the-art corporate information and computing network, which unites all KSPEU departments, which is successfully used in the educational process and scientific research [7].

1.2   Methods

An example of the successful adaptation of educational practices [8] of KSPEU for the innovative development of energy enterprises of the Republic of Tatarstan and branches of the national economy in modern socio-economic conditions is the STARTUP ENERGY project. It is aimed to creating and developing a scientific and technical association of the energy profile among young people, developing the design, entrepreneurial and creative abilities of students and their professional orientation on the real sector of the economy, energy. The profile of the association belongs to the scientific and technical industry. During the work of the scientific and technical association, students get acquainted with the methods and means of obtaining, converting, transmitting and using energy. They are taught energy and resource saving skills; students are involved in technical creativity, design and inventive activities in the field of the energy cluster, forming a professional orientation towards the energy industry as a real sector of the economy [9].

The KSPEU launched a digital career environment on the basis of the Faculty inter-university platform, which unites more than 130 universities in Russia. The system provides placement of vacancies and internships, selection of applicants, offering events, conducting tests and much more.

To meet the needs of the Russian and Vietnamese energy sectors for new generation engineers, the consortium, consisting of 3 European, 5 Russian and 2 Vietnamese universities, is implementing the project "Development of an educational program in the field of intelligent energy systems in Russian and Vietnamese universities" (ESSENCE). The project aims to modernize the existing master's programs in the field of electric power in the partner universities of Russia and Vietnam in close cooperation with industry in accordance with the requirements of the Bologna Process and the European Qualification Framework. So the programs meet all the requirements and expectations of the main stakeholders. Kazan State Power Engineering University is a partner in project "Development of an educational program in the field of intelligent energy systems in Russian and Vietnamese universities" [10].

The Electric Power Engineering Training and Research Center was established at Kazan State Power Engineering University in 2008. In 2012–2014 two training sites were built and put into operation: "Substation 110/10 kw" and "Distribution networks of RS 0.4–10 kw", during training, used for the purpose of:

- arranging for students and advanced training courses and professional retraining students to receive practical work experience at training grounds "Distribution networks of RS 0.4–10 kw", "Substation 110/10 kw»;
- get skills of operation on primary equipment and devices;
- organization, together with enterprises of the real sector of the economy, of technical training, as well as competitions of professional skills using the resource base of KSPEU;
- carrying out research work in the field of electric power;
- organization and holding of information seminars and scientific and technical conferences on energy issues with the participation of students, undergraduates, graduate students, young specialists.

The Youth Business Incubator of KSPEU ensures the development of project creativity, increased entrepreneurial literacy and business activity of students of KSPEU. It familiarizes students and CESG with the most modern methods of building a new business, including within the framework of the digital economy. Besides, it forms the conditions for the implementation of business projects for conducting educational, scientific, production, pre-diploma practices, implementation by students of research and development, taking into account the interests, proposals and orders of organizations, enterprises and companies of the real sector of the economy. It participates in the creation of conditions for the project activities of YBI residents in order to accelerate the launch of companies and their products into the market. It is illegal to consider the economic system outside social relations.

The social component, integrating into the economic system, splicing with it, becomes an integral part of it. Free competition contributes to development if it is based on an established system of social norms, and economic behavior is realized within the framework of wide values shared by society, which, according to domestic researchers is initially invested in a social matrix. There is an established system of values in society, on the basis of which all modes of economic behavior are replicated. The deeper this process proceeds, the more creative role of free competition is manifested, i.e. a serious social foundation is needed for competition [11].

Here you can pay attention to the experience of KSPEU in bachelors training in the direction of "Sociology." The uniqueness of the educational program "Economic sociology and marketing" is that it is developed taking into account the professional standard corresponding to the professional activities of graduates and based on an analysis of the requirements for professional competencies imposed on graduates in the labor market. We summarize domestic and foreign experience, consulting with leading employers, associations of industry employers in which graduates are in demand.

Graduates are highly qualified specialists who are currently in high demand in the labor market. This is due to the fact that the graduate works in research centers and organizations, departments and divisions of various enterprises, specializing in sociological, marketing and socio-economic research, public opinion research, market research, consumer behavior; territorial bodies of the Federal State Statistics Service and other statistical organizations, departments and departments, enterprises and institutions, including those specializing in the management of statistical activities.

The range of application of the formed competencies of graduates in the direction of training bachelor's degrees 39.03.01 "Sociology" of the educational program "Economic sociology and marketing" in practical activities and from the point of view of educational capital received in the process of training contributes to an increase in labor productivity in sociological and marketing departments of enterprises in the energy industry and the commercial sector, state and municipal organizations, subdivisions carrying out sociological, socio-economic and marketing research, which contributes to the achievement of their development priorities.

According to the modern requirements of the labor market and industry requests for competitive graduates of the educational program "Economic Sociology and Marketing" in the direction "Sociology," the possession of the necessary competencies for solving problems of the socio-technological type within the framework of professional activities and as a result of the application of competencies formed in the process of training within the framework of production and applied activities, bachelors of the educational program "Economic sociology and marketing" are carried out as a part of participation in the selection and adaptation of social technology for decisions of current research.

On the basis of the Department of Sociology, Political Science and Law of KSPEU of the Institute of Digital Technologies and Economics, the Laboratory of Socio-Economic Research was created. It is a training and research laboratory created to form knowledge, skills and managing and carrying out sociological, socio-economic, marketing research, "social entrepreneurship for undergoing training (introductory), production (design-technological) and production (pre-diploma) in the process of training in the direction of training 39.03.01 "Sociology" of the direction (profile) "Economic sociology and marketing." In modern socio-economic conditions, in order to identify socially mediated metrics leading to the development and success of social entrepreneurship in the youth environment, an analysis of the resource base, tool platforms and empirical research was carried out by using the method of expert assessments on the example of the Republic of Tatarstan. It implements concepts and strategies for the development of youth social entrepreneurship.

When calculating integral indicators of the effectiveness of state support for youth social entrepreneurship, weighted expert opinions and score-rating assessments were taken into account. An integral indicator of the level of socio-economic development of territorial systems should accumulate a sufficient number of private indicators [12].

The social function of entrepreneurship is expressed in reducing unemployment, increasing the number of jobs for young people, who in turn depend economically and socially on how sustainable and effective the support of state bodies of municipalities is. The social function, expressed in the ability of a young entrepreneur to show his talents and entrepreneurial initiatives, lies at the heart of the formation of a new social layer—young entrepreneurial people who gravitate to independent economic and economic activity, are able to take risks, overcome the resistance of the external environment, and achieve their goals [13].

Today, the stagnation of business entrepreneurial activity is indicated by the decreased index of growth in business activity of small and medium-sized businesses (RSBI) for the 4th quarter of 2019, but which managed to remain near 50 points. However, the modern young generation does not intend to miss entrepreneurial opportunities. So, for example, according to the RSBI support index, the business activity of "young" entrepreneurs today remains higher than the average for SMEs, including due to high personnel activity. So, for example, the indicator of youth RSBI in the 4th quarter of 2019 in the index value above 50.0 pp indicates an increase in business activity (below 50.0 pp—its decrease).

Given the active conversion nature of modern entrepreneurship, the effectiveness of the implementation of youth policy in the field of supporting the idea of   social entrepreneurship, of course, cannot be called complete based only on an impersonal metric dimension, which means that they are necessary (including taking into account retrospective changes) indicative measurements of business sentiment in the segment of youth social entrepreneurship (possibly by analogy with the calculation of the RSBI score index, but taking into account socio-oriented and regional components). Like any integral state index based on empirical indicators, the effectiveness of state support for youth social entrepreneurship requires further verification taking into account the specifics of the socio-economic development of the region (Index RSBI Support Business activity of small and medium-sized businesses. Results for 4 sq. 2019.).

2 Results

2.1 Results of an Empirical Study

Results of applied sociological research conducted in Tatarstan in 2016–2019. within the framework of the study of socio-economic activity, peculiarities of formation of life strategy trajectories of talented youth of Tatarstan made it possible to determine socio-political, socio-economic and socio-cultural metrics, which form an idea of the potential for growth and the nature of the orientation of the youth audience towards involvement in social entrepreneurship, and on the basis of which an understanding and assessment of the effectiveness of the youth policy development strategy in relation to youth social entrepreneurship in the Republic of Tatarstan can be built [14, 15].

Metrics were grouped based on indicators reflecting the centrality of the characteristic of youth as a socio-demographic group focused on the graph of social transitivity in relation to social entrepreneurship and its representatives. Among them were the following:

1. *socio-economic:* audience growth rate, which forms migration growth and outflow of youth audience; Percentage of involvement; The average degree of involvement in social entrepreneurship.
2. *socio-political:* potential for project implementation in terms of grant financing; acceptance of the terms of state investment support; perception of the level of influence of the political order in the realization of the legitimate interest of entrepreneurial activity; quantifying social tensions.
3. *sociocultural:* anticeptive viability; The success of social entrepreneurship predictions, taking into account the pace and degree of development of the regional economy; Conative orientation towards the observance of patterns of preservation of cultural conditions inherent in social entrepreneurship in the region.

Sociological researches conducted among youth (n = 1.500 people, taking into account urban and rural identities) revealed some common stereotypes formed in the social consciousness of youth (both city and village).

First of all, a group of stable social stereotypes prevails, despite socio-political bifurcation transformations, the promotion of ideas of equal opportunities and active support for the village, formulated by respondents as: the problem of finding starting conditions, condemning loved ones in connection with the instability of the essence of entrepreneurial activity, fear of economic risks, a socio-psychological barrier to taking responsibility. Secondly, the lack of a legislative and regulatory framework noted by regional researchers is the problem of regulating youth entrepreneurship, and the level of external migration, according to various estimates, is about three hundred people aged 20 to 25 years. [16].

According to official statistics, the number of youth aged 14 to 30 years at the beginning of 2019 in the Republic of Tatarstan is 781.290 people (in age groups:

from 14 to 17 years old—153.463 people: from 18 to 22 years old—179.311 people; from 23 to 30 years old—448 121 people [17].

Measurements by groups of indicators of the above metrics, such as business expectations, assessment of opportunities and readiness for the implementation of the project, availability of its investment support, tendency to master the values of social entrepreneurship indicate the specifics of the goal of the young generation in the implementation of socially significant ideas for guaranteed state support and guarantees of self-return of the project (more than 50% of respondents from among young people focused on employment in the field of social entrepreneurship). In other words, according to respondents, entrepreneurship focused on the implementation of a social mission should be economically self-sufficient, and if resources are scarce, the state should act as a guarantor of suppression, adhering to the principle of systemic support for youth social entrepreneurship.

At the same time, the model of state support for significant socio-economic projects, including in the energy sector, is based on the understanding of youth social entrepreneurship as a new way of socio-economic activity of youth, where the social purpose of project activities is accumulated with entrepreneurial innovation and achieving sustainable self-sufficiency, which determines the task of supporting social ideas and projects that can be replicated and scaled.

Adhering to the understanding of the social significance of youth social-oriented entrepreneurship as a model of business organization aimed at applying organizational innovations to solving social problems, as well as simultaneously achieving a sustainable economic effect, we are talking about innovative social projects, when the principle of systemic suppression of youth social entrepreneurship will not turn into an end in itself, but will achieve a targeted synergistic socio-economic effect.

In this regard, the result of the implemented experience of the KGEU on the formation of digital competencies in the practice of training specialists in the field of training "Economic Sociology and Marketing" seems to be relevant.

Competencies are knowledge, skills, skills, patterns of behavior and personality characteristics, with the help of which the desired results are achieved (for example, leadership, communications, etc.). The term "competence" is thought to have been first coined by Robert White [18] in 1959, who defined competence as "effective interaction (of man) with the environment."

It is known that competence is the results that determine effective work, that is, those aspects of work in which a person is competent (for example, performing the work of an accountant, sales manager, etc.). Competence implies a demonstration of skills in practice—in real working situations, and not just knowledge of theory or understanding how this is done [19].

Mastering digital competence is the basis for successful interaction with digital technologies. In general terms, digital competencies can be described as a set of knowledge and skills necessary for the use of digital technologies in activities. Digital competence is necessary for employees to perform their functions [20].

As a result of the formation of digital competency groups for students at KGEU in the training profile "Economic Sociology and Marketing," the wording of general professional competencies is reduced to the following form:

1. ability to understand the principles of modern information technologies and use them to solve the problems of professional activity
2. knowledge and understanding of the principles of modern information technologies
3. knowledge of the skills of using digital technologies to solve the tasks of professional activities, taking into account the basic requirements of information security
4. demonstration of the skills of applying correlation analysis and forecasting methods
5. the ability to apply methods of analysis, modeling, theoretical and experimental research in solving professional problems.

Also, the results of the introduction of digital competencies in the learning process are reflected in those mastered by students in the following universal competencies:

1. ability to perform search, critical analysis and synthesis of information, apply a systematic approach to solve the set tasks
2. ability to use a systematic approach to solve the set tasks,
3. knowledge of logical methods of information processing, distinguish facts from opinions, hypotheses and interpretations
4. ability to use a systematic approach to solve set tasks, to know logical methods of information processing, to distinguish facts from opinions, hypotheses and interpretations.

The result of the implemented digital professional competencies on the project type of professional tasks in the curriculum for the training of young specialists in the KGEU within the framework of the given example on the training profile "Economic Sociology and Marketing" in its wording is given as follows:

1. ability to analyze sociological and marketing research data in professional project activities
2. ability to use applied statistical data analysis programs of sociological and marketing research, web analytics
3. ability to describe the results of statistical analysis of sociological and marketing research data, development of analytical reporting.
4. ability to analyze big data using new digital technologies.

In order to analyze the peculiarities of the use of digital educational technologies in the educational process of higher education, an author's sociological study was carried out on the basis of the KGEU. As a method, a survey of students (n = 230 people, sample—target) was chosen according to the author's questionnaire, which involves block structuring of the toolkit in order to determine some indicators that mediate the influence of digital and information and educational technologies in the educational process among students. As a result of the analysis of the obtained data, it was determined that students (91%), in general, support the use of digital technologies (see Fig. 1) [21].

Digital sociology offers an alternative to the narrow definitions of digital sociological research. Some researchers define the new "computational social science" as

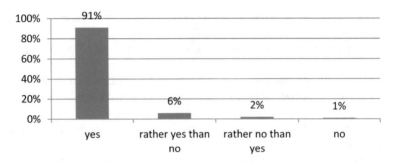

Fig. 1 Support for students using digital educational technologies,% of respondents surveyed

a form of data analysis. By contrast, "digital sociologists" seek to explore a much broader set of interactions between data, people, technology, and more that overwhelm, exceed, and do not "fit" a simple story about new forms of data analysis that require the place of old sociological research methods such as surveys or fieldwork. Digital sociology contributes to changing the role of research methods in digital society [22]. There are signs of some notable trends that are likely to take on even greater importance in the near future. First, an analysis of "digital" and everyday life. Nevertheless, a noticeable acceleration in the pace suggests that the situation will soon change. New digital technologies will inevitably open opportunities for transformational change, as in the past, with advances in "blockchain," digital verification, and quantum technologies that appear to have significant potential in this regard.

Analysis of human-technology interactions and the expected outcomes of such interactions are likely to remain relevant and will grow. The application of these new conceptual and methodological tools to understanding the diverse implications of such relationships, such as influence, on inequality and social justice, is becoming increasingly important. Finally, many discussions initiated and advanced by digital sociology, such as those concerning citizenship and identity in the digital age, look set to last for years to come [23].

The analysis of the theoretical and methodological foundations of digital sociology suggests that the systematization, advantages (or disadvantages) of existing theories and their methodologies, as well as the generalization of existing studies and their methodological provisions will be useful in the further formation of new or new content of known sociological concepts, methods, primarily aimed at experimental research and the formation of theories. Artificial intelligence, robotics, modern production automation technologies associated with the spread of the concept of the "Internet of Things" led to the fourth industrial revolution and qualitative changes in life. The stage of the birth of a "mega intellectual" society is inevitable, and it will be strengthened by the need to create digital sociological research [22].

2.2 Conclusion

Today, the specifics of the regional development of the Tatarstan economy are related to the development of territories of advanced socio-economic development (TOSER) in cities such as Naberezhnye Chelny, Nizhnekamsk, Chistopol, Zelenodolsk and Mendeleevsk, where 62 residents are registered (3 residents are at the registration stage (as of 13.11.2019)) [24]. According to experts, specifics of further development of the economy of the Republic of Tatarstan is also connected with innovations, deepening of technological base of production, formation of clusters, creation of large complexes in all socially significant and economically mediated sectors [25]. The methodological peculiarity of this approach is the use of the developed regional integrating scheme—the state model "Tatarstan $7 + 6 + 3$," which sets the logic for strategic analysis, targeting and construction of the "Strategy for the socio-economic development of the Republic of Tatarstan to 2030"—an orientation towards the growth of the region's competitiveness.

Today, the contribution of youth to the entrepreneurial environment is considered as a contribution to the development of the economic system from the perspective of the human capital of the region and the country as a whole [26, 27]. It is worth noting that youth entrepreneurship as a whole can be considered as a platform for the development of young talents in the field of business management [28, 29]. For example, the practice of forming youth associations implementing the training and collaboration of talented youth demonstrates the interest of the younger generation in social and socio-economic activity, starting with early professionalization in the field of entrepreneurial activity [30, 31]. The latter is the subject of discussion in international forums and expert meetings [32, 33].

The formation of the environment, socio-economical, political and cultural conditions for the development and implementation of youth social entrepreneurship, is certainly an integrated task of both regions and federal level. We must take into account the specifics of their socio-economic development, social entrepreneurship development potential in the region.

In accordance with paragraph 2.2. list of instructions of the President of the Republic of Tatarstan following the results of the board of the Ministry of Youth Affairs of the Republic of Tatarstan "Challenges. Priorities. Mechanisms "from 25.02.2019 No. PR-29 were amended and a new version of the state program was approved by resolutions of the Cabinet of Ministers of the Republic of Tatarstan, one of whose tasks is to manage the social development of youth, use its creative potential in strengthening the competitiveness of the republic, ensuring optimal conditions for improving the quality of life of the young generation; developing youth entrepreneurship skills.

Here, special attention should be paid to the state program, in which Kazan State Power Engineering University takes an active part—"Strategic talent management in the Republic of Tatarstan for 2015–2020," which actually became part of the developed "Strategy for the socio-economic development of Tatarstan until 2030," implemented within the framework and on the basis of the Concept for the development and implementation of the intellectual and creative potential of children and youth of the Republics of Tatarstan "Perspective," taking into account the opinion of the Association of Enterprises and Industrialists, expert opinions of the Agency of Strategic Initiatives and the Leontiev Center.

System programs of training, support, production of socially significant youth ideas are carried out under the leadership of the internal political block of the Office of the President of the Republic of Tatarstan, the Supervisory Board and the Management Board are the leading guidelines, the operator is determined by the Kazan Open University of Talents 2.0.

An important role in the work of ANO "KOUT 2.0" is played by a partner network—more than 150 organizations, including 19 centers for youth innovative and social creativity, 3 quantoria, leading universities, such as KSPEU, enterprises and public organizations. Own business projects are a form of completing the process of training young people in the field of social creativity and social entrepreneurship.

One of the platforms for the development, support and replication of current projects is the "Accelerator of youth ethno-projects"; project office of the Ministry of Youth Affairs of the Republic of Tatarstan "National-current"; Kazan Forum of Young Entrepreneurs of OIC Countries; All-Russian personnel school of youth NGOs in Kazan; Youth business incubator, etc. [34]. Youth business incubator of Kazan State Power Engineering University.

Thus, at present, advanced forms and methods of training young specialists for the energy industry of the Republic of Tatarstan and branches of the national economy in modern socio-economic conditions have been formed and actively used on the basis of the State Economic and Social Council.

References

1. Akerman, E.N.: Features of the transformation of socio-economic relations in the conditions of the development of the New Economy. Bulletin of Tomsk State Univ. Econ.. 2(14), 11–17 (2011)
2. Shafranov-Kutsev, G.F., Cherkashov, E.M.: Youth orientation on competition and entrepreneurship. Sociol. Res. 4, 117–123 (2020)
3. Yakimets, V.N., Nikovskaya, L.I.: Support of social entrepreneurship: assessment of mechanisms and rating of regions of Russia. Sociol. Res. 5, 99–10 (2019)
4. Bucci, F., Vanheule, S.: Investigating changing work and economic cultures through the lens of youth employment: a case study from a psychosocial perspective in Italy. Young. Issue 3(28), 275–293 (2020)
5. Melikhov, V.: The formation of a post-industrial type economy in Russia: the transformation of property relations in high school. Socio-economic phenomena and processes 1, 103–116 (2010)
6. Social and economic development strategies of the Republic of Tatarstan until 2030. Electronic resource, http://mert.tatarstan.ru/rus/file/pub/pub_1766090.pdf. Accessed 16 September 2022
7. Youth Innovation Center Energy, https://KSPEU.ru/Home/About/77. Accessed 1 Augist 2022

8. Garipova, R.: Special features of adaptation of young specialists at the enterprises of the energy sector: empirical results. The Turkish Online Journal of Design, Art and Communication TOJDAC Special Edition, pp. 2340–2347. TOJDAC, Turkish (2016)
9. Project STARTUP ENERGY, https://KSPEU.ru/Home/Page/65?idShablonMenu=211. Accessed 22 August 2022
10. Erasmus+ ESSENCE Project, https://KSPEU.ru/Section?idSection=8&idSectionMenu=272. Accessed 13 August 2022
11. Steblyakova, L. Transformation of economic systems: theory and practice, https://kstu.kz/wp-content/uploads/2012/12/SteblyakovaLP.pdf. Accessed 29 July 2022
12. Gainanov, D., Kantor, O., Kazakov, V.: Assessment of the level of socio-economic development of territorial systems based on metric analysis. Bulletin of Tomsk State Univ. **322**, 138–144 (2009)
13. Khizbullina, R., Muhametzanova, L., Alekseev, S.: Las perspectivas de preservación de las profesiones de trabajo entre los jóvenes en el complejo agroindustrial a la República de Tatarstán. Revista Dilemas Contemporáneos: Educación, Política y Valores: Año: VI Número: Edición Especial (66), (2018), http://www.dilemascontemporaneoseducacionpoliticayvalores. com. Accessed 12 May 2022
14. Davletshina, Ya.M., Mukhametzyanova, L.K., Khizbullina, R.R.: Monitoring of the state and the prospects of development of talented youth in modern conditions. The Turkish Online Journal of Design, Art and Communication TOJDAC (Special Edition), pp. 1562–1570 (2017)
15. Davletshina, Ya., Mukhametzyanova, L., Khizbullina, R.: Features of social well-being of modern talented youth (on research materials of the Republic of Tatarstan). Rev. Econ. Law Sociol. (4), 224–227 (2016)
16. Frolova, I.: Peculiarities of state support of youth entrepreneurship in the Republic of Tatarstan. Scientific result. Sociol. Manag. **1**(7), 57–64 (2016)
17. Age-sex composition of the population of urban districts and municipal districts of the Republic of Tatarstan: Statistical Collection of Tatarstan, https://16.rosstat.gov.ru/naselenie. Accessed 18 September 2021
18. White, R.: Motivation reconsidered: the concept of competence. Psychol **5**(66), 297–332 (1959)
19. What are competencies and why are they needed?, https://www.specialist.ru/news/1673/chto-takoe-kompetencii-i-zachem-oni-nuzhni. Accessed 14 August 2022
20. Tokareva, M.: Digital competence or digital competence. Bulletin of Shadrinsk State Pedagogical University **4**(52), 133 (2021)
21. Khizbullina, R.: Digital educational technologies in higher education: sociological aspect. Manag. Sustain. Deve. **6**(43), 72–76 (2022)
22. Pletneva, A.: Digital sociology in modern socio-economic conditions. Tinchurinsky readings—2021. Energy and digital transformation: Materials of the International Youth Scientific Conference. In 3 volumes, April 28-30. Vol. 3, p. 457-459. Kazan, Kazan State Energy University (2021)
23. Nurullina, E.: Digital sociology in modern conditions of society transformation. Society and sociology in the modern world: trends and vectors of development. IX Dylnovsky readings: Materials of the all-Russian scientific and practical conference with international participation, February 11, p. 189–195. Saratov, Saratov Source Publishing House (2022)
24. Economy of the Republic of Tatarstan, http://tatarstan.ru/about/economy.htm. Accessed 15 June 2022
25. Murakayev, I.: Peculiarities of formation of modern economy of Tatarstan. News of Tatarstan, http://www.tatarnews.ru/articles/1864. Accessed 10 December 2021
26. Jašková, D., Haviernikova, K.: The human resources as an important factor of regional development. Int. J. Bus. Soc. **21**(I2), 1464–1478 (2020)
27. Latukha, M., Veselova, A.: Talent management, absorptive capacity, and firm performance: does it work in China and Russia? Hum. Resour. Manage. **58**(I5), 503–519 (2019)
28. Muratbekova-Touron, M., Kabalina, V., Festing, M.: The phenomenon of young talent management in Russia-a context-embedded analysis. Hum. Resour. Manage. **57**(I2), 437–455 (2018)

29. Horčičková, Z., Stasiulis, N.: Philosophy of economics and management: youth participation in family business and national economy. Hum. Resour. Manage. **30**(I1), 17–26 (2019)
30. Kaliyeva, Z., Zhurasova, A., Shakhmatova, N., Uzakova, S.: Kazakhstan youth associations through the prism of public opinion of young generation. Philosophy. Sociology **30** (I.4), 325–337 (2019)
31. Hadjar, A., Niedermoser, D.W.: The role of future orientations and future life goals in achievement among secondary school students in Switzerland. J. Youth Studies **22**(I.9), 1184–1201 (2019)
32. Khayrullina, Yu., Eflova M., Garipova, R., Galieva, E.: Middle class in the republic of tatarstan: analysis of the results of sociological research. Revista San Grigorio. Special Edition, 98–103 (2018)
33. Khayrullina, Yu., Garipova, R., Tyulenev, A., Yamilov, E., Nalimova, E.: Interaction of supply chain management, entrepreneurship and consumer behaviour. Int. J. Supply Chain Manag. **9**(4), 924–930 (2020)
34. Best countries for a social entrepreneur, http://poll2019.trust.org/. Accessed 18 September 2022

The Effectiveness of the Formation and Development of In-House Knowledge in Organizations of Kazakhstan

Zoya Gelmanova, Baurzhan Bazarov, Julia Valeeva, Asel Konakbayeva, and Asia Petrovskaya

Abstract The Strategic Plan of the Republic of Kazakhstan until 2025 emphasizes the role of human capital as a factor of development in the twenty-first century: special attention is paid to the role of knowledge and education—training should be aimed not only at knowledge transfer, but also be flexible and systematic, forming appropriate competencies for future specialists, including the field of information technology, the ability to respond to changes and adapt to them. Purpose of this research is to determine the knowledge necessary for the functioning of organizational processes. The methods of this research are analysis and assessment of the current level of knowledge related to their formation and development, using the example of Kazakhstani organizations; surveys, interviewing employees and digitization by creating a virtual archive; benchmarking. In this chapter authors describe the level of participation of employees in the exchange of knowledge based on the data obtained and determine the directions for improving approaches to knowledge management. It was found that most organizations note the lack of professional competencies of their employees, including digital ones recommended by UNESCO. Organizations should pay attention to the process of training their professional employees, especially training programs related to information literacy and digital competencies, as this contributes to their professional development.

Keywords Knowledge management · Digital competences · Personnel training systems · Competency models · Information literacy

Z. Gelmanova (✉) · B. Bazarov · A. Konakbayeva · A. Petrovskaya
Karaganda Industrial University, 101400, Temirtau Republic Av. 30, Temirtau, Kazakhstan
e-mail: zoyakgiu@mail.ru

J. Valeeva
Russian University of Cooperation, Ershova, 58, 420138 Kazan, Russia

Z. Dvořáková and A. Kulachinskaya (eds.), *Digital Transformation: What is the Impact on Workers Today?*, Lecture Notes in Networks and Systems 827,
https://doi.org/10.1007/978-3-031-47694-5_10

1 Introduction

Within the setting of advance modernization of the Republic of Kazakhstan's economy, the industry of financial development, the creation of a modern economy of economic development, the industrialization of mechanical divisions and the generation of products and administrations are based on financial models that can be competitive at the rising world level. This objective in cutting edge conditions gets to be conceivable as it were in case the quality of human capital, what is moved forward and, thus, the part of information administration is reconsidered [1].

The key arrange for 2025 of the Republic of Kazakhstan emphasizes the part of human capital as an indicator within the improvement of the XXI century: uncommon consideration is paid to information and education, preparing ought to center not as it were on the exchange of information, but moreover to be more adaptable, systemic, form modern important competencies within the future master, and be able to reply changes and make alterations [2].

The victory of any cutting edge organization depends to a degree on its capacity to utilize information and advantage from assets. Be that as it may, the changing headcount in this respect postures challenges for organizations. A noteworthy number of experienced workers are resigning, taking part-time employments or taking off their employments. This comes about in a misfortune of corporate memory. The catalysts of the issue are calls for fetched investment funds to keep the current beneficial workforce ahead of the saves in utilize by workflow optimization. The later financial downturn due to the widespread has caused a decay in numerous organizations, coming about in a misfortune of information.

The results of such issues for organizations incorporate misfortune of proficiency, time, openings to attain vital objectives, decreased levels of representative and asset satisfaction, and expensive costs of attempting to exchange lost knowledge, which eventually leads to a potential decrease within the viability of the organization.

Hence, the foremost significant issues are the arrangement and change of information in organizations of present day advertise conditions in Kazakhstan for the development and change of unused sorts of corporate societies and framework, comparing to the fundamental standards of information administration.

2 Methods

Nowadays, the existing models of instruction are primarily frameworks that meet the prerequisites of an mechanical society and are pointed at obsolete ways of learning that don't meet the necessities of the twenty-first century. Only a couple of created nations of the world since the 1990s. of the final century, endeavors started to develop unused frameworks based on the information economy, where there's a move in emphasis on the mass generation of information.

Any changes in advancement patterns within the instruction framework of the XXI century ought to be related with free get to and make an opportunity for the spread of human thoughts and shrewdness, and ought to moreover be reliable with the arrangement of financial, political and natural issues [3].

All the comes about of this think about were gotten within the handle of information investigation amid a overview of 9 Kazakhstani organizations. All the data gotten was carefully examined, digitized by making a virtual chronicle, and checked for input mistakes. All information gotten from quantitative inquire about from organizations was scrambled for fair and subjective examination.

The overview of respondents was carried out employing a study of respondents. For answers, a five-point Likert scale from 1 to 5 (from total dissent to outright understanding) was received. The likelihood of the nearness of respondents who don't have a supposition, or who want to acknowledge a position of nonpartisanship, was too taken into consideration—for this, the reply "I will abstain from answering" was entered within the reply choices. In arrange to conduct a subjective investigate, all the collected information were analyzed utilizing the strategy of substance examination.

In arrange to preserve secrecy and secrecy, the names of all nine organizations were coded utilizing letters of the Latin letter set. The letter doled out to each organization has nothing to do with the title of the company.

The destinations of this information collection apparatus were as takes after: to decide the nearness of frameworks that direct the forms of administration and the arrangement of in-house information in organizations; decide the accessibility and level of in-house preparing for workers.

The survey comprised of three segments: statistic information, evaluations of the activities of organizations within the arrangement and administration of information.

3 Results

Over the final decades of the XXI century, the intrigued and consideration of both organizations and the complete world is progressively centered on the arrangement, advancement and administration of information.

Progressively, inquire about is being conducted to evaluate existing strategies, frameworks and tools for information administration, but nowadays there's no all inclusive framework for surveying the comes about of the arrangement and administration of information [4].

The approach to making and directing preparing and capacity advancement frameworks inside organizations can contrast essentially depending on the characteristics of the circle of action of a specific organization, in any case, by the by, there are a number of features characteristic of overseeing the potential of workers of organizations completely different nations—they can be conditionally separated into two Categories: Eastern (Japanese) and Western (European-American) approach.

The Japanese approach to employee potential management implies division into two main categories: employee training "on the job"—that is, directly during work;

and training the employee "outside of work"—i.e. additional training provided by firms outside of working hours. Modern Japanese companies also distinguish a third type—self-education, considering that it is self-education and training in the workplace that are the main directions in the system of knowledge management and personnel potential; however, it should be noted that in this case, additional measures are used to improve qualifications and develop potential—such as short-term refresher courses for employees, depending on their position, specialty and, accordingly, depending on emerging topics and problems necessary for their further professional development. A separate feature that distinguishes this approach from others is that the need to develop capacity management systems and, accordingly, personnel training is enshrined at the legislative level, since Japan has adopted a law on professional training of personnel. Thus, the development of specialists as professionals and the improvement of their qualifications are almost entirely under the responsibility of Japanese companies, and government agencies and institutions play only a subsidiary role. Thus, Japanese organizations receive specialists who have the professional knowledge, skills and abilities necessary specifically for their activities, ensure staff consistency of the team and contribute to the creation, preservation and development of not only explicit, but also implicit knowledge: the main principle of their development systems is the process of transformation "knowledge of one employee into knowledge of the entire organization."

The general view of the system of training, development and retraining of the organization's specialists consists of five main components: basic training, vocational training, retraining in relation to improving the quality of specialists' abilities, advanced training, training of instructors-mentors [5, 6].

The Western approach to the advancement of the potential of pros is basically based on a double framework—a combination of hypothesis and hone, i.e. advancement of an representative without "interference" from proficient exercises: for illustration, in Germany and Switzerland it may be a clearly structured framework of staff advancement, taking into consideration specialized and mechanical outside changes (counting those anticipated), changes related to the exercises of the organization itself, as well as changes, related to the necessities for the calling of the master himself, in this manner giving preparing for the master for changes related with his exercises (counting the usage of unused proficient assignments). At the same time, the proficient improvement of an employee's potential can be carried out not only on the premise of the organization's capabilities, but moreover with the assistance of "commonwealth networks" of a few organizations [7, 8].

It ought to be famous that in differentiate to the Japanese approach, where there's a clear line of partition between common instruction and professional instruction (where the most proficient improvement is given by in-house instruction), the European-American preparing framework expect that the representative as of now has shaped competencies that require a brief adjustment period to begin working. The frameworks of in-house preparing of masters can be characterized as the "Ford-Taylor concept", that's, profoundly specialized improvement based on expanding the competence and proficient advancement of a worker with respect to a particular assignment and a particular work [9].

In this way, it is conceivable to decide the headings of these two approaches to the framework of capacity building and work force preparing as:

growing the capabilities of pros in arrange to guarantee their portability—Japanese, even framework;
proficient advancement of pros inside the system of a particular proficient action of a master—European-American, vertical framework.

Frameworks for the improvement of the potential of representatives of organizations are a component of the work force administration framework and speak to a number of exercises based on staff arrangement and pointed at preparing, progressed preparing or retraining of organizational work force—these forms are a handle pointed at expanding the potential of a specialist's proficient movement, his financial and social comes about, as well as expanding the human capital of the organization, its levelheaded utilize and expanding the effectiveness of exercises in common [10].

The framework for expanding the potential of the organization's workers can be conditionally partitioned into two fundamental categories: specifically, in-house preparing of an worker (the method of exchanging and making, counting verifiable information) and outside preparing, i.e. expanding the polished skill of a pro outside his "work" exercises (the method of getting express information).

The method of expanding the potential of a pro from the side of in-house preparing incorporates strategies such as turn, appointment, mentoring, and a strategy of complicating errands. Off-the-job preparing strategies incorporate trainings, addresses, refresher courses.

In-house preparing is straightforwardly related to the commonsense component of a specialist's action—in this way, this sort of preparing permits a representative to make strides his proficient level in connection to certain errands, relate his work with the objectives of the organization's exercises, and more clearly get it his position and objectives as a master in a given organization. Be that as it may, when executing the forms of in-house preparing, there are a number of challenges, among which are the pertinence of information exchanged to the worker, the pertinence of the preparing program itself, the level of readiness of a pro who plays the role of a guide, and his devotion.

The in-house preparing framework could be a complex handle comprising of coherently interconnected stages: examination and evaluation of the require for faculty advancement, the arrangement of a support for the arrangement of the preparing prepare, the definition of preparing objectives and goals, the planning of the program and the choice of strategies of in-house training, and, at long last, the appraisal of the gotten/information, abilities and capacities of a master changed/improved within the learning handle. It is important to note that not only mentors and workers ought to take portion within the forms, but moreover supervisors—everybody ought to be curious about the method and realize the complete require for such occasions [11, 12].

Training outside the organization contributes to the development of a specialist as a professional in terms of acquiring a new type of knowledge, skills and abilities, as well as a new behavioral line, thereby contributing to the introduction of new ideas and knowledge into the organizational structure of the organization and its activities.

All factors affecting the development of employees can be conditionally divided into three categories: external (the level of socio-economic development of the state, the educational system, changes in the legislative and legal system of the country, the state of work market, the emergence of competitors), internal (material and technical base of the organization, working conditions, labor incentives, the presence of motivational measures, management style of managers and the characteristics of the organization's team).

Managing the development of an organization's employees involves analyzing and assessing the professional capabilities of a specialist (performance by an employee of his professional duties and work functions, direct assessment of employee competencies, as well as training, retraining and advanced training of a specialist), but also the creation of all kinds of conditions that provide a certain level of comfort for a specialist and contributing to an increase in the level of professionalism and labor efficiency of a specialist—creating the labor necessary for normal activity, motivating an employee (including through a system of remuneration, bonuses and other incentives, creating an atmosphere conducive to development, supporting employees in terms of career, taking into account the interests and needs of specialists), maintaining the climate both in the team and in the organization (relationships in the team, the relationship of the team with the leader).

Thus, from the point of view of logical functioning and ensuring the highest possible efficiency, the personnel development assessment program can be formulated as an assessment of personnel development along three dimensions (economic, professional and social benefits of employees) on the basis of qualitative, quantitative and complex methods of analysis.

The process of internal training and development of personnel is not only a tool for solving certain problems and tasks of the organization, but also an important resource for unlocking the potential of human capital of the organization, an additional source of knowledge formation. A well-thought-out, flexible and balanced program of professional development of the organization's specialists will increase the employee's motivation, ensure and increase his loyalty to the organization, ensure the increase of professionalism of the organization's specialists—thus, the organization's personnel development management system, through continuous improvement and updating of specialists' knowledge and increasing their efficiency, will ensure and contribute to the organization's competitiveness in the field of its activities.

In modern market conditions, from the organizational point of view, the knowledge and skills of specialists have become one of the most important factors in the success of a company.

Education is one of the most important and indispensable components of social development and personal development of modern specialists. Education makes it possible to correct existing deficiencies and reduce gaps, including those that have emerged in the course of the development of world civilization under the influence of digitalization, automation, transformation of social institutions and demographic changes.

One of the main goals of education is the acquisition of a common set of knowledge, skills, abilities and other competencies for a fundamental understanding of the subject under study, which in turn is necessary for self-realization and the development of society as a whole.

Today, there are several models of education that are being prioritized and actively developed: personalized learning, lifelong learning.

Any change in the modern education system in the twenty-first century must be linked to free access, create opportunities for the dissemination of human ideas and wisdom, and be consistent with the solution of economic, political and environmental problems. Buttles and Stodinger argue that a new social model based on "collective intelligence" must be created to play an important role in decision-making and human development [3].

The main task of education in preparing modern experts is to develop their professional competence. Professional competence is a generalized concept encompassing a set of competencies that ensure that an expert meets the requirements of his or her profession.

The competencies of today's modern specialist include not only the ability to harness and use "hard" professional skills, but also so-called "flexible" competencies (e.g., communication, systems analysis, information literacy, creativity, responsiveness, critical thinking, and multidisciplinarity), as well as the formation and development of personal qualities for one's self-fulfillment and sustainable development of the whole of humankind. Thus, in the report "Education for the Future", Luksha and Afanasiev proposed a four-level model of competencies (Fig. 1) [13].

As shown in Fig. 1, this competence life cycle model has the following levels: narrowly defined professional competences, so-called "interdisciplinary" cross-contextual competences, meta-competences (which refer above all to cognitive, logical, physical, moral-emotional and other types of thinking), and existential competences, which are more basic concepts of knowledge, skills and abilities. The length of time it takes to master a particular type of thinking varies from a few months to a lifetime. Contextual competencies are easy to learn and quick to correct, but have one significant drawback-they quickly become outdated and lose relevance. In contrast, cross-situational competencies can remain relevant for decades, but mastery of such skills and knowledge takes more time. Meta-competencies and survival competencies, despite the complexity of their development, are the most important competencies in the modern world and are designed to meet the needs of changing economic and social systems.

There are many models for the development of capabilities, but today there is no universally accepted model that fully and completely captures and categorizes all necessary and relevant capabilities.

UNESCO also noted the need for a quality control system in the field of education and the associated processes for the formation and continuous improvement of competencies. Each country has its own view of the organizational chart of core competencies. For example, Singapore lists social skills, thinking skills, information skills, creativity skills, collaboration skills, knowledge application skills,

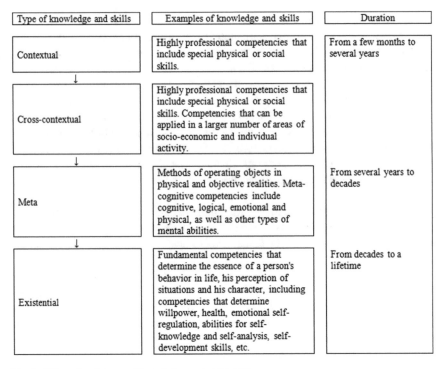

Fig. 1 Life cycle of types of knowledge and skills [13]

literacy skills, self-improvement skills and personal development skills as essential competencies [14].

New Zealand identified five key qualities: using language skills, self-organization, communication, thinking. Thinking skills, building interpersonal relationships, active participation, self-organization skills, and competent use of symbolic and symbolic (textual and linguistic) data.

Australia has adopted a ten-competency model: qualifications, thinking skills, self-improvement, teamwork skills, socio-ethical skills, ability to work and apply ICT products, international understanding, creativity, and quantitative thinking skills.

According to Indonesia, basic competencies can be considered as knowledge, intelligence, personality traits, self-education skills and self-regulation.

The Harvard Center for Curriculum Redesign (CCR), led by Feydl, with the support of the Organization for Economic Cooperation and Development (OECD), has created a unique model of an integrated educational organization that can define relevant competencies for the twenty-first century. The main idea of this model is to create a new space conducive to self-learning and making personal decisions about one's future. This model allows for the definition of goals and provides a common basis for the redesign of existing educational models, and is also characterized by the enhancement of transformative competencies based on information about the relevance of a specific type of knowledge. The model of the CPUP reflects the interplay between the four dimensions within the framework of a four-dimensional model of education: in addition to the traditional skills, knowledge, and character, a new face emerges—meta-cognition—which is an internal comprehension and self-adaptation process of personal learning. internal understanding and self-adaptation processes. The reason why metacognition needs to be emphasized as a separate facet is that metacognition improves the process of using competencies in domains beyond what is generally accepted.

Metacognition can be defined as a thinking process in which the current state of the educational process is analyzed, goals are identified, behavioral strategies are formed, and program outcomes are predicted.

According to Hacker and Danlosky, the metacognitive process can be conditionally divided into 3 verbalization levels: knowledge transformation, non-verbal knowledge transformation, and knowledge interpretation transformation. Metacognition is developed in the context of the student's current task, and regardless of its initial level, metacognition improves the student's mastery of knowledge, skills and abilities [15].

Therefore, learners who are confident in their ability to achieve their goals are more likely to use metacognitive practices, thereby improving the quality of learning and increasing the likelihood of increased productivity and efficiency. Not only theoretical activities, but also practical activities, combined with methods and techniques, contribute to the development of knowledge, skills and abilities of trainees and to the development of competent, conscientious and judgmental specialists.

Modern educational learning models should aim at creating solutions to problems and developing further learning, including by forming and integrating competences and their requirements related to all stages of the education system, ensuring logical relations and continuity; at the same time, the model itself should be formed not only by compiling and analyzing the requirements of the State, but also of the labour market and society as a whole, taking into account not only highly specialized competencies, but also individual, multidisciplinary and generic ones.

Modern post-industrial societies need to revisit the relationship between factors of production and human resources. The effectiveness of knowledge management has a direct impact on the productivity and performance of modern organizations. Low human capabilities lead to lower quality of products and services, reduced competitiveness, increased costs and deterioration of social and labor relations. Therefore, the role of qualified human resources capable of managing and coordinating production processes as drivers of organizational activity is increasing.

Despite continuing advances in science and technology and major leaps forward in the field of information technology, organizations of all sizes are faced with a common dilemma: how to meet the continuing demand for high-quality talent in the face of a shortage of talent in the labour market. The lack of highly qualified people with the skills to use the latest technological and societal achievements is a constant problem for business leaders.

According to the data obtained from 9 Kazakhstani organizations, the distribution of employees of organizations by level of education revealed a prevailing difference between holders of a bachelor's degree (total percentage—65.85%) and holders of secondary general education (30.49%). At the same time, out of 108 respondents with a bachelor's degree, 24 have a diploma with honors, a master's degree—5 people (3.05%), and only 1 person has a certificate of general secondary education.

The largest number of specialists with a bachelor's degree work in organizations D (14.81%), E (12.96%) and I (12.04%), the smallest—7.41%—in firm F.

According to the results of the second survey, the respondents' work experience in organizations ranges from 1 to 23 years: 17 worked less than or about 2 years (10.37%), 26—3 years (15.85%), 18 people—4 years (10, 98%), 5 years—19.51% (32 employees), 11 people—about 8 years (6.71%), 1.83% was the number of those who have been working in the company for 16–19 years, more than 20 years 2 people work (1.22%). As it follows from this, most of the employees have been working in the organization for 2 to 7 years. The average length of service of the respondents was 5.98 years, which testifies to the relative stability in staff turnover among the administrative and management personnel.

The issue of general work experience was also studied, taking into account the experience of work in other companies in similar fields of activity in similar positions. One year was noted by 0.61%, fourteen people worked for about 2 years (8.56%), from 3 to 5 years was indicated by 10.98% (18 people), 35 employees indicated 6–9 years (21.34%), 10–11 years old—38 people (23.17%), 24 specialists marked 12–13 years old, 19–25 years old—3.05%, 25 years and more—nine people (5.49%).

Despite the relative stability among the workforce, the majority of respondents indicated a desire to change jobs or dissatisfaction related to their position (the exceptions are companies B and C, where there is an increased turnover of personnel that exceeds the permissible norm and is about 10–12%). The reasons were such factors as the lack of incentives in the form of career growth or an increase in wages. According to a survey conducted among 164 employees of administrative and managerial personnel of 9 organizations operating in various fields of activity (from industrial to trade), the majority of respondents (53.05%) are not satisfied with the possibilities of additional, specialized training and advanced training.

The data obtained in the course of the survey indicate that the majority of respondents are somehow familiar with such concepts as "knowledge", "qualifications", "competence", "personnel management" (about 84.14% of respondents).

The majority of managers, according to the research results, are not familiar with the concept of knowledge management (30.43%), have some ideas—56.53%, or are familiar exclusively in the theoretical aspect (13.04%). Among all employees of the administrative and management personnel (managers of all levels and accountants took part in the survey), the level of awareness of the concept of knowledge management is also not high enough—acquaintances of varying degrees of awareness of the concept—70.12, 20.73% are completely unfamiliar with this concept The formation of external and internal sources of knowledge is an important component in knowledge management processes. The main sources of external information can be conditionally divided into general, institutional and market ones, while internal ones are represented by knowledge and information created in the course of the organization's activities. According to the results of a survey among the leaders of Kazakhstani companies, the main sources were the consumer market, Internet resources and information obtained from partners and in the course of analyzing the activities of competitors. The extremely low level of institutional sources should be noted: the percentage of those who noted this type of resource turned out to be extremely low—6%. The internal sources of knowledge are the results of the work of marketing, production, as well as economic (in some cases) departments of organizations.

It should also be noted that there is a rather low percentage of use of various specialized systems, for example, customer relationship systems, specialized training programs for employees—only one of nine organizations had such a program, but it is also rarely used in practice.

About 35% of managers consider knowledge management to be economically unprofitable, or do not consider it important, 17.39% of managers are not sure of its effectiveness, and more than 45% consider this direction of management to be promising. It should be noted that among the respondents who spoke negatively about the prospects of knowledge management (KM), most of them are top managers.

According to the results of the study, about 30.18% of managers are dissatisfied with the qualifications of their employees regarding their positions, 26.34% refrained from commenting, 43.48% are to varying degrees satisfied with the work of their employees.

About 75% of the respondents agreed with the statement that various educational activities have a positive effect on the effectiveness of the organization's activities.

Despite the fact that according to numerous studies, the obsolescence of knowledge is on average about 20%, while the recommended period for acquiring new knowledge in the industrial sphere (in particular, metallurgy) is every 3–4 years, and in the business sphere, the period is reduced to 2—3 years old. In particular, with regard to representatives of the financial department of organizations—accountants, adjustments are made annually to the databases concerning taxation and other legislative acts, the knowledge of which is necessary for the correct and productive work of a specialist.

Despite the recognition of training as a factor that directly affects the efficiency and success of the company, in most organizations there is no investment in the education of specialists. The main reason is that the costs are not economically viable (according to senior management).

There were questions that were devoted to the provision of the necessary computers and software, as well as their relevance in relation to modern conditions. More than 60% believe that organizations are fully provided with all the necessary hardware and software. It should be noted that some respondents noted the need to update the existing equipment—14.02%.

The value of information technology as a tool necessary for the correct and effective operation of the enterprise was noted by more than 83% of the respondents, only 3% considered the products of information and communication technologies meaningless.

According to the data provided by 9 organizations, only 3 companies are engaged in advanced training of administrative and management personnel. Issues related to the presence in organizations of an official knowledge management system and the presence or absence of a document confirming the existence of a knowledge management policy showed the absence of such in organizations (A—90.91%, B—86.67%, C—94, 74%, D—82.61%, E—85%, F—85.71%, G—75%, H and I—88.24% and 94.44%, respectively).

The comparative characteristics of which indicate the constancy of the ongoing activities dedicated to professional development and the constancy among employees who undergo training at the expense of the organization, however, it should also be emphasized that these activities were rather isolated. During the survey, it was also noted that many employees would like to take any refresher courses (or they go on their own), but organizations do not have the ability or desire to do so. Thus, the question arises about the need to analyze the situations occurring in these organizations to assess the presence or absence of the fact of inconsistency in the competencies (including knowledge) of employees.

As noted earlier, competencies are a significant factor in HR management. Competence is a set of personality traits and behavioral patterns, as well as knowledge, skills and abilities used by an individual, in particular, in his professional activity. Some organizations use competencies as their primary HR tool.

Competencies, according to Hitt and Huskisson, are the totality of the resources and capabilities of the enterprise. At the same time, the so-called key competencies become a valuable source of competitive and strategic advantage and at the same time play the role of functions and current processes of the company.

Competencies can be classified into business organizational competencies, which are expressed in the structure and processes of organizational activities and do not depend on the individual employee, and individual competencies, which belong to the individual employee and reflect the employee's knowledge, skills, experience, and level of skill sufficient to carry out the occupational activity effectively, as well as ethical and personal attitudes.

The strategic management literature also categorizes core and unique competencies. Core professional competencies represent collective learning in an organization, focusing on management tasks such as diversification of various job-specific skills and combining multiple streams of information simultaneously. These competencies should be maintained and improved through accumulated experience, as the more productive the organization is, the more emphasis it places on the educational process, and thus the faster it reduces its own cost levels.

Core competencies represent the internal potential of an organization and are the basis of its strategy, competitiveness and profitability, while distinctive competencies represent the ability to improve the competitiveness of the organization.

To achieve the organization's vision and mission, the core competencies must match the strategic objectives.

Traditionally, many organizations have dealt with personnel management issues using, to a greater or lesser extent, key competencies and professional competencies, or simply individual competencies to assess employees among managers.

Since organizations do not have a clearly developed system of assessing the professional competences of administrative and managerial staff, we have analysed the requirements set by the heads of organizations when recruiting administrative and managerial staff.

Since this field of activity is represented mainly by representatives of the so-called "brain work", based on the results of the survey of managers and staff members themselves, the following competencies were identified as necessary for specialists: professional skills, strategic and critical thinking, communication, ICT skills, result orientation, managerial skills.

Of course, each occupation has its own specificities, but after an overall consideration of the competencies of AUP specialists, all responding organizations arrived at this set of competencies.

Since the object of the study is directly administrative and managerial personnel, and since the activities of the organizations are different (from industrial organizations to organizations in the field of trade), a generic competence model was created, which takes into account managerial, special and basic competences of professionals in the economic professions (Table 1).

In accordance with this competency model, a survey was conducted among employees of the administrative and managerial personnel of 3 organizations involved in improving the qualifications and competence of personnel. It should

Table 1 Model of key competencies of employees of administrative and managerial personnel of organizations

Competencies	Characteristic
Result orientation	Ability to effectively implement assigned tasks. Availability of skills that ensure the quality of work
Customer focus	The ability to identify, analyze and provide the customer with the necessary product/service, while focusing both on the interests of the organization and on the needs of the customer
Strategic and critical thinking	The ability to holistically see the situation, analyze factors, both external and internal, that affect the success and effectiveness of an organization, department or individual, both in the medium and long term, as well as the ability to analyze the data obtained (in particular, information), on their basis, form your own vision, which serves as a guide to further actions and decision-making
Communication and sociability	Ability to establish effective working relationships, as well as interact in other social, cultural and other contexts
ICT—literacy	Basic knowledge of digital technologies and their products (including knowledge and ability to use a personal computer and its main, as well as specialized programs)
Management skills	Ability to evaluate, plan and make decisions, taking into account all possible factors that can affect the processes in the course of the organization's activities. When holding a managerial position—also be able to analyze and regulate the activities of subordinates for their effective work
Professional skills	Availability of knowledge, abilities, skills directly related to the performance of official duties

also be noted that since the representatives of this unit are the heads of organizations, middle and operational managers, as well as accountants of organizations. Due to the difference in the required competencies, it was decided to separate the results of the representatives of these two specialties. For the assessment of competencies, the following gradation was adopted—a 4-point system—basic, intermediate, advanced, professional. The survey took into account the opinions of managers, employees themselves, their immediate superiors and colleagues. The results of organizations A, E and I are presented in Figs. 2, 3, and 4, respectively.

The analysis of the provided results showed that, on average, all indicators are within the advanced, 3 levels. At the same time, the data indicate that several indicators of the competence of the organization's employees coincide with the necessary ones. For employees of the accounting department, these are competencies related to customer focus, communication and management skills (in organizations A and I).

Among managers of organization E, indicators of professional skills and ICT competencies coincide with the required level, and in company I—the level of

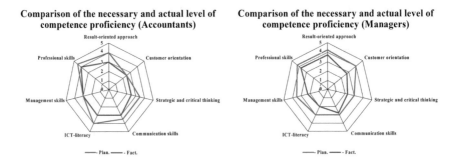

Fig. 2 Comparative characteristics of the required and actual level of competence of specialists in the administrative and managerial personnel of the organization I

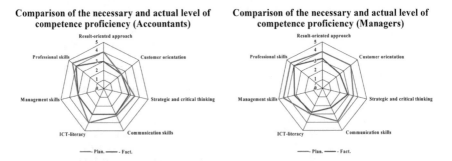

Fig. 3 Comparative characteristics of the required and actual level of competence of specialists in the administrative and managerial personnel of the organization E

Comparison of the necessary and actual level of competence proficiency (Accountants)	Comparison of the necessary and actual level of competence proficiency (Managers)
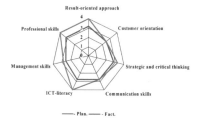	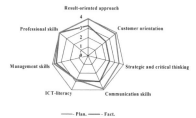

Fig. 4 Comparative characteristics of the required and actual level of competence of specialists in the administrative and managerial personnel of the organization A

communication. The greatest deviation in all three organizations in terms of "customer focus", as well as critical and strategic thinking.

ICT literacy showed the greatest deviation from the required skills—in all three organizations. According to the results obtained in the questionnaire survey, similar problems were reflected in the organizations not considered in this article 6. The main reasons were named such factors as lack of awareness and lack of opportunities for advanced training, employees themselves either do not have motivation or financial capabilities. Today, special attention should be paid to the development of this type of competence.

4 Discussion

A large-scale study conducted in American companies showed that a 10% increase in the cost of personnel training increased labor productivity by 8.5%, while the same increase in capital investment increased labor productivity by only 3.8%. At the same time, in-house professional training increases labor productivity. At the same time, in-house training of professionals increases labor productivity and ensures revenue growth. Research shows that every year of extension of employee training leads to an additional 3% increase in GDP. The most developed internal training systems are considered to be those of such countries as Japan, the United States, France and Korea.

The use of the above-mentioned methods by educational institutions and organizations in developing training programmes for specialists at all levels will ensure that competencies are in line with the modern requirements of employers and the labour market, and will also help to develop human resources with modern interdisciplinary competencies, which, in turn, will help to reduce the level of shortage of qualified personnel in the Republic of Kazakhstan.

5 Conclusion

The main goal of the State Program for the Development of Education and Science in the Republic of Kazakhstan for 2020–2025 is to increase the competitiveness of education and science in Kazakhstan, to develop and train individuals on the basis of universal values. Thus, knowledge becomes the center of economic transformation of Kazakhstan, the most important source of well-being and the key to maintaining the competitiveness of individuals, organizations and socio-economic development of the country [1–3].

Modern education should meet all modern conditions and be aimed at completely new ways of learning that would meet the requirements of the twenty-first century: training of a specialist should be aimed at developing modern, relevant professional competencies, which are a set of competencies (not only traditional competencies, but also new ones, including interdisciplinarity and meta-cognitive skills), corresponding to the requirements of his specialty.

Despite the awareness of the importance of personnel management, the considered organizations do not implement and do not use knowledge management systems and tools, which significantly reduces their potential. According to the results of the study, none of the organizations under consideration has systems similar to them. At the moment, all processes related to knowledge management processes are a decentralized process in separate departments, occurring exclusively within certain areas of knowledge. However, the creation of such a system, or at least the use of some of its tools, is absolutely necessary, especially in relation to organizations B and C: the use of some elements of knowledge management will reduce the loss of implicit knowledge, increase the level of competence of their employees, and partially reduce tension in the team (since clearly spelled out instructions will facilitate the transfer of information and some of the conflicts will disappear), which in turn will increase efficiency in organizations.

Taking into account the data obtained in the course of the analysis, organizations need to develop a well-defined system of internal training adapted to modern conditions. High-quality internal training system will create and improve the necessary conditions for continuous professional growth of employees, provide initiative for improving their skills, provide opportunities for professional and personal growth, reduce turnover (staff turnover) and, ultimately, optimize expenditures. In particular, this will raise the level of qualification of specialists, increase the motivation of employees, and additional activities related to it will contribute to cohesion within the team, thus preventing possible problems (for Company A and Company I), partially solving existing problems related to the qualification of specialists (for Company B and Company C), creating relative independence on the labor market, and contributing to the improvement of the organization's production of final quality of products/services produced by the organization.

Therefore, we can conclude that in order to improve efficiency and performance, organizations need to develop their own up-to-date personnel training systems, taking into account all the needs of the organization and its employees. The internal training system should take into account all aspects: from the improvement of professional and background skills to the development of personal qualities (e.g., leadership, initiative, improvement of entrepreneurship and other business qualities). As a means of improving the quality of personnel, it is possible to provide professional training at institutions of higher learning, attend professional courses, acquire additional specializations, retraining courses, attend seminars, rotations, and self-study.

The lack of a clearly formulated competence model prevents the management of organizations from objectively assessing the competence of professionals, which is reflected, in particular, in the system of remuneration and bonuses, which, in turn, reduces the efficiency and effectiveness of the activities of professionals and directly affects the atmosphere of the team and the activities of the organization as a whole. Many leaders note the inadequacy of existing critical and creative thinking and competencies.

Most organizations have noted the lack of certain competencies, both professional and general (e.g., critical thinking and creative skills), among their employees, but they themselves do not contribute to the development and training of these competencies [14].

For effective personnel management, organizations need to take into account the specifics of knowledge management, including attention to the process of training and education of personnel, which implies: creating and using an organizational culture that is used, inter alia, for the training and evaluation of personnel; and cooperating with various educational institutions on the education of specialists, combining the process of education with the development of professional skills and the careers of professionals. Organizations should pay attention to the process of training of specialists, in particular to training programmes and their costs, as this contributes not only to the professional development of specialists, but also to the maintenance and improvement of their motivation, as well as to the maintenance of their emotional, physical, mental and cognitive health.

References

1. The Library of the First President of the Republic of Kazakhstan, https://presidentlib.kz/en/news/third-modernization-kazakhstan-key-formula-development-state-and-ensure-its-competitiveness, Last accessed 18 Nov 2022
2. The Food and Agriculture Organization (FAO). https://www.fao.org/faolex/results/details/en/c/LEX-FAOC191496/#:~:text=The%20strategic%20goal%20of%20the,human%20capital%2C%20technological%20modernization%2C%20improving, Last accessed 18 Nov 2022
3. Strategic Development Plan of the Republic of Kazakhstan until 2025. https://policy.thinkbluedata.com/node/4025, Last accessed 18 Nov 2022
4. Baltes, P.B., Staudinger, U.M.: Wisdom: a metaheuristic (pragmatic) to orchestrate mind and virtue toward excellence. Am. Psychol. **55**, 122–136 (2000). https://doi.org/10.1037/0003-066X.55.1.122,lastaccessed2022/11/18
5. Gelmanova, Z.S.: Assessment of key competencies of metallurgical production workers. Int. J Appl Fundam Res (9-2), 101–105 (2014). https://applied-research.ru/ru/article/view?id=5841, Last accessed 18 Nov 2022

6. Nonaka, I., Takeuchi, H.: The knowledge-creating company: how Japanese companies create the dynamics of innovation. **284** (1995). https://papers.ssrn.com/sol3/papers.cfm?abstract_id= 1496713, Last accessed 18 Nov 2022

7. van Zolingen, S.J.: Developments in education and training in Japan. (2005). https://eric.ed. gov/?id=ED492317, Last accessed 18 Nov 2022

8. Pilz, M.: The future of vocational education and training in a changing world. Education and training: issues, concerns and prospects (2012). https://www.researchgate.net/publication/291 758882_The_Future_of_Vocational_Education_and_Training_in_a_Changing_World, Last accessed 18 Nov 2022

9. Rebmann, K., Tenfelde, W, Schlömer, T.: Berufs-und Wirtschaftspädagogik: Eine Einführung in Strukturbegriffe, Gabler Verlag; 4. Aufl. 2011 edition, German 268 (2011)

10. Fedorov, M.V.: Strategic management human resources, management of economic systems. Electron. Sci. J **11**(59), 29 (2013). https://cyberleninka.ru/article/n/strategicheskoe-upravlenie-chelovecheskimi-resursami-2, Last accessed 18 Nov 2022

11. Gelmanova, Z.S., Spanova, B.Z., Kudaibergen, B.E., Silaeva, T.O.: Formation of creative education as the basis of personality development. Int. J. Appl. Fundam. Res. (4-3), 572–575 (2017). https://applied-research.ru/ru/article/view?id=11517 (Accessed 25 07/25/2023), Last accessed 11 Nov 2022

12. Gelmanova, Z.S.: Assessment of key competences of metallurgical production workers. Int. J. Appl. Fundam. Res. **9**(2), 101–105 (2014)

13. Trunkina, L.V., Krumina, K.V.: Labor Potential management: information and communication Aspect. Russian entrepreneurship, No. 17 (2012). https://cyberleninka.ru/article/n/upravlenie-trudovym-potentsialom-informatsionno-kommunikatsionnyy-aspekt, Last accessed 18 Nov 2022

14. Gelmanova, Z.S.: Organization of vocational training at work. Int. J. Exp. Educ. **8**, 17–21 (2016). https://expeducation.ru/ru/article/view?id=10345, Last accessed 18 Nov 2022

15. Douglas, J., Hacker Dunlosky, J., Graesser, A.C.: Metacognition in Educational Theory and Practice. New York (1998)

Speech Technologies in the Operational Personnel Training of Urban Rail Transport Systems

Ekaterina P. Balakina⬚, Maksim A. Kulagin⬚, Ludmila N. Loginova, and Valentina G. Sidorenko

Abstract The process of technical re-equipment of railway transport facilities is inextricably linked to the transformation of technological processes subject to automation and to the complexity of microprocessor systems of railway automation. The range of reports on non-standard situations arising during the operation of technological processes, the results of diagnostics of units and elements of microprocessor systems, etc., is being expanded. The speech recognition issues for operational personnel training of urban rail transport systems in order to automate the learning process and assess the trainee's skills are considered. As an example the metro is used. To process audio messages between a train dispatcher and other workers and services various speech recognition methodologies are applied. The approach to the dispatcher commands classification is proposed. The generating commands algorithm for a control system under limited amount data conditions is synthesized. The approaches review to the text construction processing and speech recognition systems is carried out. The intelligent system block diagram for recognizing the voice dispatcher commands is given. The continuous speech recognition quality is assessed.

Keywords Intelligent control · Neural networks · Speech recognition · Deep learning · Urban rail transport · Subway · Operating personnel training

1 Introduction

Modern society is characterized by an increased demand for the electronic computers use in all areas from education to health care, as well as in various man-machine systems, which include automated/automatic control systems, on-board

E. P. Balakina (✉) · M. A. Kulagin · L. N. Loginova · V. G. Sidorenko
Department of Control and Information Security, Russian University of Transport, Moscow, Russia
e-mail: balakina_e@list.ru

Sirius University of Science and Technology, Sochi, Russia

hardware systems and queuing systems. Such systems are actively used in the traffic management processes in ensuring the transport systems objects safety.

Multi-tasking systems are being developed and deployed worldwide, with a focus on user interface to ensure efficient systems use with maximum convenience. Modern society is aimed at increasing the interaction efficiency between the operator and the computing system using the most natural methods for a person to exchange information—gestures, writing and speech [1]. The entering information method from touch panels has found its application, however, the most attractive for use in such systems and the most promising is the voice interface, because it is the most natural and stable communication way [1].

The world's largest companies in their development offer a speech interface as an alternative to a graphical one. Moreover, in special-purpose systems, voice notification operator systems the about various events have been used for a long time to improve the feedback efficiency. In this view, to speed up the operator's work in a special-purpose system, increase the reliability and his work efficiency, it is advisable to use direct communication between the operator and the computer through voice and, as a result, obtain a full-fledged speech interface.

In 2019, Microsoft and Nuance began developing tools for automating administrative work in medical institutions, in particular, documentation. Nuance Communications is well known for developing speech recognition technologies. These technologies were used in the virtual assistant Siri, and now the company focuses primarily on the medical industry. The company serves 77% of US hospitals with medical records and imaging facilities [2]. The car manufacturer Hyundai Motor uses voice recognition technology to control the multimedia system using voice commands to ensure system safe operation while driving [3]. The companies group "CRT" LLC is developing a software solution for creating and maintaining phono records, conducting automatic person identification by voice, based on such automatic research voice and speech methods, which don't depend the language, speaker's accent, dialect and text content [4].

Speech recognition is the main component of the voice control subsystem. Currently, end-2-end neural network architectures based on sequence-to-sequence learning (seq2seq) [5, 6] are attractive for automatic speech recognition, because they can study both acoustic and linguistic information in unlike conventional ASR (Automatic Speech Recognition) systems, which were based on hybrid hidden Markov models and deep neural network models. Moreover, end-2-end models are suitable for compression because they do not need separate phonetic dictionaries and language models, making them one of the best candidates for mobile applications. There are already formed solutions for continuous speech recognition [7].

Most models for speech recognition and feature extraction from speech use Mel Frequency Cepstral Coefficients (MFCC), extracted from continuous speech [8]. In this case, the ASR system performance is determined using the word error rate (WER).

Named Entity Recognition (NER) refers to the identifying task said designations from text that belong to predefined semantic types such as person, location, organization, etc. NER serves as the basis for many natural language applications such as answering questions, synthesizing text, and machine translation. In recent years, deep learning, supported by continuous vector representations with real values and semantic composition through nonlinear processing, has been used in NER systems, providing high performance [9]. Most NER systems use methods based on linguistic grammar [10], statistical models, and machine learning methods [11]. NER statistical systems typically require a large manually amount annotated training data.

Research in the detection and commands recognition in text and speech has been carried out in various fields. For example, in [12], algorithms and approaches to the dialogue systems construction in a smart home are described. A model for organizing local (without using the Internet) speech recognition using mobile devices was proposed in [13]. Chapter [14] is devoted to the voice control subsystem creation for an intelligent motion control system of autonomous mobile robot. The work [14] proposes the voice control subsystem structure based on the CMU Sphinx speech recognition system.

The foreign works number are devoted to the speech recognition systems development and research [15, 16]. Thus, in [15], the authors propose a basic model structure for adapting the speech recognition system to multi-domain operation and preventing thematic errors.

2 Methods

The voice control use is gaining particular relevance in transport systems. The traffic control quality in transport, including urban rail transport systems (URTS), is subject to increased requirements to ensure the passenger transportation safety and convenience [17, 18]. In this regard, special attention is paid to the operational personnel qualifications involved in traffic control. Various conducting training sessions methods are used to train and improve the qualifications of the URTS operating personnel. For example, the Train Dispatcher Simulator is used on the Moscow metro, which, through the metro line multifunctional model, makes it possible to bring the training process as a real control process [19, 20] (see Fig. 1).

Conducting training sessions takes place in real time and takes a certain time amount, while both the trainer and the trainee give voice commands and duplicate them into the model, interacting with line model using the keyboard and mouse through the pop-up menu system. The trainer selects a planned training situation and simulates messages from line workers. The trainee receives voice messages from the trainer and reacts to them by giving voice commands. All voice commands from the trainee are duplicated by the trainer in the simulator as commands entered to the services involved in the organizing the movement process.

The integrated approach use in training and advanced operating personnel training is relevant [17], which is primarily based on the wide range solution of automation issues at all levels using the training and knowledge assessment system for operating personnel, namely:

Fig. 1 The training process functional diagram of URTS operating personnel

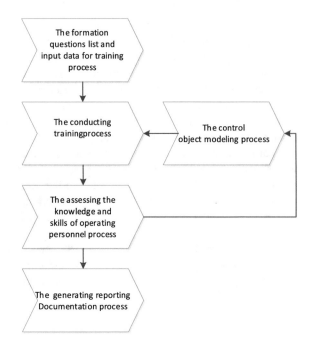

in the setting the initial data for the training session process;

in the conducting a training lesson process;

in the assessing knowledge and skills in organizing the operational personnel movement process;

in the obtaining reporting documentation process.

The voice recognition subsystem introduction into the training tools will significantly facilitate the interaction process between the trainer and the trainee with the Integrated Automated Training and Knowledge Assessment System of the URTS operating personnel (hereinafter referred to as the Integrated System).

The voice recognition subsystem introduction into the training tools will significantly facilitate the interaction process between the trainer and the trainee with the Integrated Automated Training and Knowledge Assessment System of the URTS operating personnel (hereinafter referred to as the Integrated System).

The following requirements are imposed to the voice recognition subsystem:

- phrases recognition, their independent construction;
- informing about the submitted command;
- sending the command for execution to the Integrated System;
- voice commands logging in text and audio files. The protocol should provide for the both registration dispatcher's and "electronic" commands and the instructor's commands;
- electronic orders automatic filling;

- equipping the dispatcher's workplaces with floor pedals to enable communication "to transmit" for the simplex information transmission system organization. At present, the workstations of the DCH are equipped with a pedal with an "interruption", when the "line sounds" are turned off when the pedal is pressed. In the subsystem, when the pedal is pressed, the "electronic" workers responses should not sound; it is necessary that the workers "wait" for the dispatcher to release the pedal;
- supplementing the instructor's workplace with signaling the voice command execution;
- the ability to disable the dispatcher's voice command execution by the instructor;
- the information list commands for the trainee to familiarize with it before the training session is provided.

As research part carried out in this chapter, an open model for Russian language recognition was used, which includes hidden Markov models, a mixture model of Gaussian distributions, deep neural networks, namely Time-Delay Neural Networks (TDNN) [21]. This model was chosen on the comparative analysis results basis carried out in the chapter [22], as the model that shows the highest recognition accuracy, slightly losing to competitors in speed terms. This model and the Kaldi speech recognition system built on its basis are more suitable for scientific research than their counterparts [23].

This study purpose is to automate the controlling process vehicles movement during the training URTS operational workers by automated dispatcher's voice commands recognition to exclude mechanical interaction with the control panel, which will lead to a decrease in the student errors number.

Within the research it is necessary to solve the following tasks:

To analyze the existing speech and text methods processing in recognition, commands classification and named entities extraction from the text.

To develop a text classifier by the message type in it.

To generate a dispatcher commands classifier.

To develop an algorithm for classifying dispatcher commands based on text processing.

To analyze the proposed algorithm work.

To develop intelligent dispatcher voice command recognition system structural diagram included as a voice recognition subsystem in the Integrated Automated Training and Knowledge Assessment System of the URTS operating personnel.

The main requirement for the system being developed is the accuracy and audio-to-text conversion module reliability, as well as the breaking the text into replicas of the dispatcher and other employees accuracy. Each replica goes for processing by other system modules, which fully rely on the text classification high accuracy. The system includes modules for classifying text by command type, searching for commands in the text, and searching for subjects and stations in the text.

The Intelligent Voice Command Recognition System structural diagram includes elements that implement the following actions (see Fig. 2):

Fig.2 Intelligent system block diagram for recognizing the dispatcher's voice commands

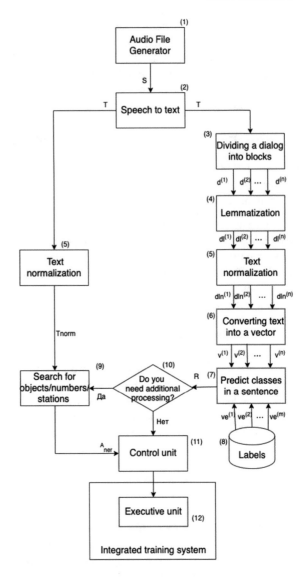

Audio files Generator—recording the dispatcher and other workers conversations in the *S* file, which is then transferred to the system input for processing.

Convert speech to text—converts the audio signal to text format *T*.

Dividing a dialog into blocks—splitting files into n possible sentences that are used to classify $[d^{(1)}, d^{(2)}, \ldots, d^{(n)}]$.

Lemmatization—words conversion to the standard form $[d_l^{(1)}, d_l^{(2)}, \ldots, d_l^{(n)}]$.

Text normalization—converting numeric entities into text, removing uninformative words $[d_{ln}^{(1)}, d_{ln}^{(2)}, \ldots, d_{ln}^{(n)}]$.

Converting text into a vector—representing the sentence words in the length n sequences space (embedding) $[v^{(1)}, v^{(2)}, \ldots, v^{(n)}]$.

Text classification by the type message in it—each n sentences assignment to one of the possible classes: "Informational", "Command", "Order", «Undefined». Vector classes calculation R (possible commands)—metric calculation between the input text and the reference examples commands vector $[v_e^{(1)}, v_e^{(2)}, \ldots, v_e^{(m)}]$.

Knowledge base of commands reference types for organizing search in a sentence.

Search for objects/number/stations in the text—based on information about the command class and the need for additional search for action objects, the search for objects in the text is carried out up to the current considered sentence T_{norm}.

Checking the additional message processing is required depending on the message class. If the message entails the command execution automation, the command data is sent to the control unit.

1. Control unit—this block accepts the commands array for execution as input and transmits them to the executive block.
2. Execution unit—this unit receives data from the control unit as input and is responsible for their execution and is a direct part of the Integrated automated training and assessing the knowledge system of the URTS operating personnel.

3 Results

In this chapter the results research of the algorithm for the sentences classification and the search for commands in the text are presented.

To solve the sentences classification problem, a convolutional neural network was used. The neural network input is a 36×300 matrix, where 36 is the tokens (words) number in the sentence (text depth), and 300 is the embedding vector depth for each word. The word transformation into a numerical vector is carried out on the neural network use pre-trained on the literary text corpus (corpus size is more than 150 GB) [24]. The text depth was chosen based on the 0.9-quantile of the words number statistical distribution in a sentence in the collected dataset (see Fig. 3).

In this chapter the results research of the algorithm for the sentences classification and the search for commands in the text are presented.

To solve the sentences classification problem, a convolutional neural network was used. The neural network input is a 36×300 matrix, where 36 is the tokens (words) number in the sentence (text depth), and 300 is the embedding vector depth for each word. The word transformation into a numerical vector is carried out on the neural network use pre-trained on the literary text corpus (corpus size is more than 150 GB) [24]. The text depth was chosen based on the 0.9-quantile of the words number statistical distribution in a sentence in the collected dataset (see Fig. 3).

As the training result neural network for the sentences classification during 100 training iterations, a forecast accuracy at about 94–96% was achieved on the test sample (see Fig. 4). This result allows us to accurately determine the text class.

Fig. 3 The words distribution histogram in the text

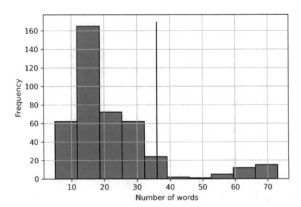

The examples of classified sentences (the probability belonging distribution vector to classes contains the probabilities values in the following sequence: "Informational", "Command", "Order"):

"On the first main line of the Airport-Sokol haul, the voltage was removed from the contact rail"—[0.99,0.01,0];

"Train number one hundred and one route number three, while following the Airport-Sokol section, the overload relay to Sokol station was triggered to disembark passengers and proceed to the Sokol depot"—[0.0.942.0.058];

"Order number eighty station Sokol train driver number four hundred and one route number eleven to follow when traffic lights are prohibited until the permissive indication appears"—[0, 0.013, 0.987].

If the classifier assigns the text to the "Command" class, then the algorithm for searching for commands in the text is launched.

The algorithm consists the following steps:

A reference commands possible types classifier is formed (Table 1), in which each word in a command is represented as a numerical vector based on a pre-trained neural network.

For each word in the text, a numerical vector is calculated based on the pretrained neural network.

Each reference command "sliding window" goes through the text and calculates the distance between the reference vector matrix and the analyzed text section using the metric:

$$r_{ij} = \frac{v^{(i)} \cdot v_e^{(j)}}{\|v^{(i)}\| \times \|v_e^{(j)}\|} = \frac{\sum_{k=1}^{L} v_k^{(i)} \times v_{ek}^{(j)}}{\sqrt{\sum_{k=1}^{L} \left(v_k^{(i)}\right)^2} \times \sqrt{\sum_{k=1}^{L} \left(v_{ek}^{(i)}\right)^2}} \tag{1}$$

$$R = \begin{bmatrix} r_{11} & \cdots & r_{1m} \\ \vdots & \ddots & \vdots \\ r_{n1} & \cdots & r_{nm} \end{bmatrix} \geq t,$$

Fig.4 Results training
neural networks

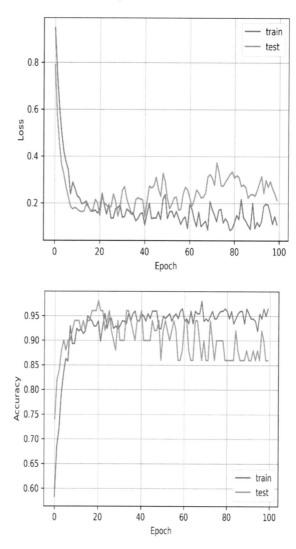

where r_{ij} is the cosine distance between the main text embedding vector and the reference command; t is the threshold at which the search result for the reference command in the text is determined.

If the distance between the reference text and the window is less than the specified threshold t (within the current study, the threshold $t = 0.5$ was used), then the system provides information that the command was found in the text.

Table 1 Reference
commands

№	Command	Team group
1	Don't go from the station	Traffic ban
2	Wait, don't move	Traffic ban
3	Stay away from the station	Traffic ban
4	Stay, wait for the command	Traffic ban
5	Don't go without a command	Traffic ban
6	Follow turnover	Traffic permit
7	Follow at intervals	Traffic permit
8	Follow the reserve to the depot	Traffic permit
9	Leave the station at intervals	Traffic permit
10	Follow to the depot	Traffic permit
11	Go in reserve	Traffic permit
12	Go at …	Traffic permit
13	Go at intervals …	Traffic permit
14	Change cockpit	Other commands
15	Make the delay	Other commands
16	Drop off passengers	Other commands
17	Prepare a reserve squad	Other commands

4 Discussion

We will assess the proposed continuous speech recognition method quality according to the following indicators:

- correctly recognized commands share;
- erroneously recognized commands share;
- unrecognized commands share.

As the research result the following quality indicators of continuous speech recognition were obtained: 98.3% of correctly recognized commands, 0.5% of erroneously recognized commands, 1.2% of unrecognized commands.

The results are acceptable for use to automate the conducting operational personnel training process of the URTS.

5 Conclusions

The main chapter results:

- the literature review was carried out on solving the text classification problem and analysis, speech recognition and building intelligent dialogue systems;

- developed an approach to solving the text classification problem by command type;
- the reference teams classifier has been developed;
- proposed algorithm for commands searching in the text based on the metric distance between the reference commands and the source text;
- trained convolutional neural network for text classification by command type;
- the continuous speech recognition results were obtained and the recognition quality was assessed according to the selected indicators.

In the future it is planned to test the developed algorithms work on an increased data amount, develop methods and algorithms for solving the problem of finding named entities in the text, determine the subject and object, and develop an algorithm for determining the location on which the action is carried out.

Acknowledgements The reported study was funded by RFBR, Sirius University of Science and Technology, JSC Russian Railways and Educational Fund "Talent and success", project number 20-37-51001

References

1. Alekseev, I.V., Mitrokhin, M.A.: Modern speech recognition methods for constructing a voice-control interface for special purpose systems. News of higher educational institutions. Volga region. Tech. Sci. **2**(50), 3–10 (2019)
2. Microsoft buys speech recognition technology developer Nuance Communications. https://www.tssonline.ru/news/microsoft-pokupayet-rasrabotchika-tehnologiy-raspoznavaniya-rechi-nuance-communiations. Last accessed 09 Aug 2021
3. Voice recognition system. http://webmanual.hyundai.com/STD_GEN5_WIDE/AVNT/EU/Russian/voicerecognitionsystem.html. Last accessed 09 Aug 2021
4. Matveev, Yu.N.: Technologies for biometric identification of a person by voice and other modalities. Eng. J. Sci. Innov. **3**(3), 46–61 (2012)
5. Battenberg, E.: Exploring neural transducers for end-to-end speech recognition. In: Conference 2017, IEEE, Automatic Speech Recognition and Understanding Workshop (ASRU), pp. 206–213. IEEE (2017)
6. Prabhavalkar, R., Rao, K., Sainath, T.N., Li, B., Johnson, L., Jaitly, N.: A comparison of sequence-to-sequence models for speech recognition. Proc. Interspeech 939–943 (2017)
7. Guglani, J., Mishra, A.N.: Continuous Punjabi speech recognition model based on Kaldi ASR toolkit. Int. J. Speech Technol. **21**, 211–216 (2018)
8. Huang, X., Acero, A., Hon, H.-W.: Spoken Language Processing: A Guide to Theory, Algorithm, and System Development. Prentice Hall PTR, p. 980 (2001)
9. Li, J., Sun, A., Han, J., Li, C.: A survey on deep learning for named entity recognition. IEEE Trans. Knowl. Data Eng. **34**(1), 50–70 (2020). https://doi.org/10.1109/TKDE.2020.2981314
10. Dashtipour, K., et al.: Persian named entity recognition. In: 2017 IEEE 16th International Conference on Cognitive Informatics and Cognitive Computing (ICCI* CC), pp. 79–83. IEEE (2017)
11. Mansouri, A., Affendey, L.S., Mamat, A.: Named entity recognition approaches. Int. J. Comput. Sci. Netw. Secur. **8**, 339–344 (2008)
12. Eremenko, V.O., Molodyakov, S.A.: Smart home control using dialogue commands of voice control. Modern Technologies in the Theory and Practice of Programming, pp. 17–18. (2020)

13. Shmatkov, V.N., Bonkovski, P., Medvedev, D.S., Korzukhin, S.V., Golendukhin, D.V., Spynu, S.F., Muromtsev, D.I.: Interaction with devices of the Internet of things using voice interface. Sci. Tech. Bull. Inf. Technol. Mech. Opt. **19**, 714–721 (2019)
14. Toporin, A.A.: Voice control subsystem of the intelligent control system of a mobile robot. Bull. Sci. Educ. **14–4**(92), 9–13 (2020)
15. Moon, J., Yun, S., Lee, D., Kim, S.: A preliminary study on topical model for multi-domain speech recognition via word embedding vector. In: 2019 34th International Technical Conference on Circuits/Systems. Computers and Communications (ITC-CSCC), pp. 1–4. (2019)
16. Bouafif, L., Ouni, K.: A speech tool software for signal processing applications. In: 2012 6th International Conference on Sciences of Electronics, pp. 788–791. Technologies of Information and Telecommunications (SETIT) (2012)
17. Baranov, L.A., Sidorenko, V.G., Balakina, E.P., Loginova, L.N.: Intelligent centralized traffic management of a rapid transit system under heavy traffic. Dependability **21**, 17–23 (2021)
18. Baranov, L.A., Sidorenko, V.G., Balakina, E.P., Loginova, L.N.: Integrated approach in the training of operational workers of urban rail transport systems. Transp. Sci. Technol. **2**, 22–31 (2021)
19. Baranov, L.A., Balakina, E.P., Erofeev, E.V., Sidorenko, V.G.: Multifunctional models of control systems. News of Higher Educational Institutions. Problems of printing and publishing, vol. 2. pp. 79–82. (2012)
20. Baranov, L.A., Balakina, E.P.: Prospect for multifunction models. World Transp. **10**, 70–74 (2012)
21. The Kaldi speech recognition toolkit. https://www.researchgate.net/publication/228828379_The_Kaldi_speech_recognition_toolkit. Last accessed 09 Aug 2021
22. Belenko M.V., Balakshin P.V: Comparative analysis of open source speech recognition systems. Int. Sci. Res. J. **4–4**(58), 13–18 (2017)
23. Markovnikov, N.M., Kipyatkova, I.S.: Analytical review of integrated speech recognition systems. Trudy SPIIRAN **3**(58), 77–110 (2018)
24. Kutuzov, A., Kuzmenko, E.: WebVectors: a toolkit for building web interfaces for vector semantic models. In: International Conference on Analysis of Images, Social Networks and Texts, pp. 155–161. (2016)

Smart Workplace: Implementation of Management Innovations in Public Administration Processes

Marina Ivanova⬤, Grigory Kulkaev⬤, and Natalia Mozaleva⬤

Abstract The chapter examines issues of the effectiveness of introducing management innovations into public administration processes in order to modernize the workplace and space of civil servants. A theoretical basis is being formed on the concepts of management innovation and digital technologies and government. It is concluded that such innovative implementations are ineffective. Based on this, the authors propose an algorithm consisting of 8 steps, directed. Within the framework of this algorithm, a unified model of the functioning of digital technologies in the process of public administration is developed, it is proposed to introduce the "client path" method to analyze the satisfaction of actors with the functioning of processes, and it is also proposed to classify management innovations and digital technologies depending on the scope of application and the ultimate goal for the convenience of choosing certain technologies for specific purposes. Thus, the proposed algorithm is aimed at increasing the efficiency and effectiveness of introducing management innovations into public administration processes in order to simplify, automate and speed up the work of civil servants.

Keywords Management innovation · Digital government · Smart workplace · Public administration · Digital technologies · Public service

1 Introduction

Introducing management innovations into public administration processes can have significant benefits for efficiency, effectiveness, and overall performance. By leveraging workplace technologies, public managers can enhance their ability to fulfill their responsibilities and meet the needs of both their colleagues and citizens.

Workplace technologies have evolved to encompass not only office tools but also instruments for social interactions and knowledge exchange among colleagues

M. Ivanova · G. Kulkaev (✉) · N. Mozaleva
Peter the Great St. Petersburg Polytechnic University, Saint Petersburg, Russian Federation
e-mail: kulkaev_g@spbstu.ru

© The Author(s), under exclusive license to Springer Nature Switzerland AG 2023
Z. Dvořáková and A. Kulachinskaya (eds.), *Digital Transformation: What is the Impact on Workers Today?*, Lecture Notes in Networks and Systems 827,
https://doi.org/10.1007/978-3-031-47694-5_12

as well as citizens. These technologies enable public administrators to facilitate information exchanges and engage with stakeholders on various platforms. With the advent of readily available intelligent technologies and big data, public administration is undergoing constant reconfiguration of its digital and human interface. This reconfiguration is necessary to keep up with the changing demands of the digital age and enable public administrators to effectively harness the potential of these technologies. The emergence of a smart workplace in public administration signifies a shift towards a more modern and efficient state [1].

In the 1990s, there was a push to modernize the state and make public administration more efficient through the use of new Information and Communication Technologies [2]. These technologies, particularly those related to the Internet, brought about significant changes in public administration. Public administrators were able to leverage these technologies to streamline processes, improve communication, and enhance service delivery to citizens. In recent years, the importance of sophisticated technologies such as the internet of things, sensor systems, big data analytics, and artificial intelligence has become increasingly apparent in public administration. These technologies have the potential to transform the way public administration operates by establishing new service delivery models that integrate physical, digital, public, and private environments. Furthermore, these smart government initiatives go beyond traditional digitization endeavors by rethinking the relationship between public administration and its stakeholders. By using new technologies, public administration is seeking to reimagine its interactions with stakeholders and create new ways of delivering services. These initiatives aim to create a more connected and integrated public sector that utilizes advanced technologies to improve efficiency, effectiveness, and responsiveness.

Thus, the chapter proposes to analyze the issue of management innovations in the public administration system, as well as digital technologies that are being introduced into management processes within the framework of these innovations. Based on the analysis, the chapter will propose an algorithm for introducing management innovations into public administration processes, present a classification of these innovations, as well as a unified model for implementing the public administration process.

2 Management Innovation and Digital Government

Managing any enterprise, organization or public authority is a rather complex process. In the context of digital transformation, this process becomes more complicated due to the increasing speed of changes in external conditions and the need for digital technologies. As a key method of stimulating development in these conditions, the most relevant method is the introduction of innovations.

Innovation as an object of management has been identified in post-industrial society. At previous stages of the development of society, innovation was not considered as one of the factors of competitive success, and accordingly was not singled out as a separate subject of research and management [3].

In a broad sense, innovation is defined as the transformation of potential scientific and technological progress into real progress, embodied in new products and technologies. Innovation can be defined as an innovation in the field of engineering, technology, labor organization or management, based on the use of scientific achievements and best practices.

The chapter will consider managerial, or organizational, innovations. Today, management innovations are least studied by scientists; the attention of scientists and practitioners is increasingly drawn to production, technical, technological innovations that contribute to the rapid and obvious acquisition of competitive advantages by the management object. But it is worth noting that increased productivity and high economic performance are achieved by those organizations that consistently implement management innovations [4]. Innovations related to management ensure an effective transition from the initial state of the reformed organization to the desired one, and the updated management system makes it possible to implement technological innovations.

A striking example of organizational innovation in the field of state and municipal management is the implementation of the concept of e-government. Electronic government in many countries has reduced the burden on state and municipal employees and simplified the processes of interaction between authorities between departments, citizens and businesses.

The main result of the introduction of such management innovation was the concept of providing state and municipal services through a "one-stop shop". One window is a term that refers to a technology for providing services to citizens and businesses. The "one window" technology aims to reduce the time of forced communication between citizens and businesses and is characterized by the fact that the provision of any services is concentrated in one place, from filing an application to issuing the results of a decision of an executive or other body [5].

Today, the main impetus for the implementation of management innovations in the field of state and municipal government is the implementation of the concept of digital government and the introduction of its technologies into the work of executive authorities.

The introduction of digital government technologies as management innovations helps to reduce the number of state and municipal employees, automate and simplify processes within government bodies, as well as increase the effectiveness and efficiency of their activities.

In the scientific literature, the terms electronic and digital government are often equated. However, they are still different.

The term e-government (EG) is understood as an updated model of the activities of state authorities, local governments, as well as other organizations involved in the implementation of the powers of authorities, using information and communication technologies to implement their functions [6].

Digital government is the next stage of digitalization of public authority after e-government, the concept of which is to use the advantages of digital data and technologies to create, optimize and transform the activities of government bodies.

Various technologies are used in the formation of digital government. Today there are 5 digital government technologies:

1. Blockchain.
2. Big data.
3. Cloud services.
4. Artificial intelligence.
5. Internet of Things.

So, digital government, having specific features at its core, can neither be identified nor characterized as part of e-government. Digital government is an independent concept, which is built on the basis of the achievements of e-government, but differs from it in a number of technological, structural and conceptual characteristics [7]. Digital government is seen as a means to improve the efficiency of the state as a whole, including both service delivery and decision-making. Having a high level of flexibility and long-term potential, digital government can fundamentally change the existing system, raising public administration to a qualitatively new level.

3 Results: Algorithm for Introducing Management Innovations into Public Administration Processes

Despite the active policies of many countries to introduce digital technologies into public administration processes, not many countries have seen significant changes in the public administration system. For example, the introduction of management innovations and the introduction of digital technologies are designed to increase the efficiency of public administration by reducing the number of civil servants, automating processes and simplifying them. In some countries, on the contrary, after the introduction of management innovations and digitalization of power, an increase in the number of civil servants, an increase in the degree of bureaucratization of processes and an increase in the time frame for their implementation are noticed [8].

To solve this problem, it is proposed to develop an algorithm for the introduction of digital government technologies as management innovations for public authorities.

Despite the positive effects from the introduction of individual digital government technologies, the implementation of digital government technologies still needs to be considered comprehensively, together with each other, to achieve the maximum positive effect. Figure 1 shows a unified block diagram of the functioning of digital government technologies in the process of state and municipal management.

The introduction of these processes into the general processes taking place in executive authorities will help save time and budget funds, simplify the work of civil servants, which in turn will serve as the basis for rebuilding human resources to suit the digital conditions of the modern world.

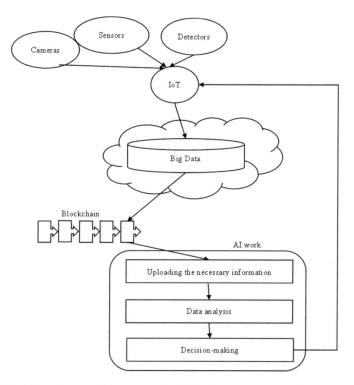

Fig. 1 Unified block diagram of the functioning of digital government technologies in the system of state and municipal government

Thus, having determined the main place of digital government technologies in the processes of state and municipal government, it is possible to develop an algorithm for the implementation of management innovations during the digitalization of public authority.

Step 1: Building a process flowchart.

Building a flowchart is necessary to visualize any of the processes of state and municipal government. The construction of such models contributes to the decomposition of the stages of the process, which will subsequently make it possible to analyze this process. Figure 2 shows a primitive process flow diagram.

By breaking the process into several stages, you can identify problematic issues using the next step.

Step 2: Defining the "customer journey".

Since management innovations in state and municipal administration are aimed at improving the quality and efficiency of public authority, it is necessary to identify target actors for further changes. As mentioned earlier, the role of actors can be citizens, organizations (external) and state and municipal employees (internal). For

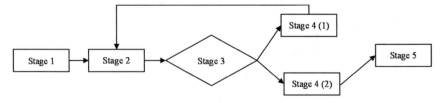

Fig. 2 Process flow diagram

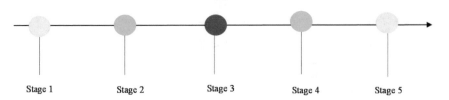

Fig. 3 "Customer journey"

everyone, the essence of the "client journey" is one thing: a description of the client's (actor's) emotions at each stage of a process (Fig. 3).

In the figure, red denotes negative emotions and impressions from the implementation of a process stage, green—positive, yellow—the average between negative and positive. Thus, building a customer journey helps to identify the strengths and weaknesses of the process, the latter of which are subsequently subject to organizational changes.

Step 3: Identifying problematic stages.

After comparing the impressions of the actors from the implementation of the stages, it is necessary to identify problem areas, that is, what is the problem of the actor at a particular stage of the process.

Among the problem areas are the complexity of the stage, the high probability of making an error, duration, etc.

Step 4: Define digital government technology.

To determine the digital government technology that will solve the identified problem, it is necessary to classify the organizational changes that arise during the implementation of management innovations through the introduction of digital government technologies (Table 1).

This classification gives an understanding of, firstly, what organizational changes are possible as a result of management innovations, and secondly, what organizational changes correspond to individual digital government technologies.

Thus, returning to the algorithm, the developed classification will contribute to the correct choice of digital government technology, depending on the identified problems in the processes and the necessary changes in order to improve the activities of state and municipal government bodies.

Table 1 Classification of organizational changes as a result of management innovations through the introduction of digital government technologies

Type of classification	Type of organizational change	Digital government technologies involved in change
In the area of organizational change	Changes in the final product, result Changes in the organizational structure of the authority Changes in state and municipal management processes	Big data, cloud services, blockchain, IoT, AI
By actors of organizational change	External (interaction with citizens, organizations) Internal (interaction between state (municipal) employees)	Big data, cloud services, blockchain, IoT, AI
According to organizational changes in the structure	Reduction of structural units and their number of personnel Creation of new structural divisions	Big data, cloud services, blockchain, IoT, AI
On organizational changes in processes	Automation Simplification Increased security Acceleration Reducing stages	AI, IoT AI, IoT, Big Data, cloud services Cloud services, blockchain Cloud services, blockchain, IoT, AI IoT, AI

Step 5: Development of an updated process flow diagram taking into account the technologies being introduced.

At this stage, a new flowchart of the process is being developed, which already reflects the place and role of digital government technologies in the process. For example, Fig. 4 shows an updated flowchart from Fig. 2 using artificial intelligence at the branching stage of the process.

Thus, the introduction of AI in this case made it possible, first of all, to automate and speed up the decision-making process at stage 3, as well as reduce the total number of stages.

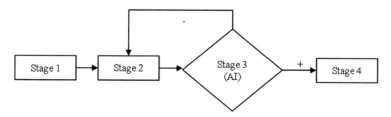

Fig. 4 Updated flowchart using artificial intelligence

Step 6: Determine the approach to manage the project.

Every change to an established model or structure is a project. And many of them fail because of the wrong management tools. In some places it is necessary to follow the "waterfall" model, and in others, in Agile/Scrum. Moreover, modern realities lead us to hybrid approaches, when different approaches are used within a project at different stages.

You should remember the main stages of both any project and any stage within a project:

1. Initiation.
2. Planning.
3. Implementation.
4. Control.
5. Completion.

The "Entangled Projects" method from the Kinevin model, which works on the principle of "research-realize-react," is well suited for implementing such projects.

Step 7: Launch of a pilot project and adjustments.

This stage involves testing the project and the operation of the process. Typically, pilot projects are launched either in a specific territory or in a separate executive authority.

If the pilot project is unsuccessful, it is necessary to return to step 1 and carry out work to improve the system again.

Step 8: Consolidate the results.

If positive results emerge from the implementation of the pilot project, it is necessary to extend this experience to other regions or authorities to improve the selected process throughout.

Based on the results of the work done, you should also consolidate the results obtained: analyze what was done well and where there were failures. Based on the results of the pilot implementation, instructions, memos, regulations are developed, personnel are trained, and restructuring occurs (if necessary).

4   Discussion

Despite the unified approach to the development of the algorithm, the results of this study still cannot be applied to absolutely all processes of public administration. Many authors argue, in addition to the fact that each state has its own characteristics of public administration that are distinctive from other countries, but within each state there are peculiarities of the functioning of public administration processes, depending on the field of activity, level of management, etc. [9–12].

However, most authors point out that for the widespread improvement of public administration processes as part of the introduction of management innovations and digital technologies, it is necessary to develop a general methodology for the implementation of such innovations [13–15]. Based on the developed general methodology, each state will be able to build its own internal policy to create smart jobs and processes for civil servants. Therefore, based on this provision, we can draw a conclusion about the practical significance of the results obtained.

5 Conclusion

Digitalization of state and municipal government is a complex process due to the need for clear legal regulation, established bureaucratic traditions and the scale of the state apparatus. Nevertheless, without the digitalization of public power, the digital transformation of other areas will not take place, since the state is the center of society.

This chapter presented the results of a study on the topic of a smart workplace for a civil servant, which was ensured by the introduction of digital technologies into operational processes as management innovations. The article examines the theoretical aspects of management innovation and digital government, substantiates the connection between them and the need for their implementation, and lists digital technologies through which the restructuring of management processes occurs.

After analyzing the scientific literature on the topic, it was concluded that the introduction of management innovations into the public administration system is ineffective. As part of solving this problem, an algorithm for introducing organizational innovations was developed and proposed. A special feature of this algorithm was the development of a unified model of the functioning of digital technologies in the process of public administration, the introduction of the "customer path" method in process analysis, as well as the development of a classification of management innovations that facilitates the selection of the necessary digital technology to improve the process.

As part of promising directions for further research, the authors plan to conduct a statistical analysis of indicators of the effectiveness of public administration and the impact of the implementation of the previously proposed algorithm on these indicators. In addition, the authors plan in future work to present a specific action of the developed algorithm to demonstrate the effectiveness of its work.

References

1. How Local Government Is Leveraging Technology to Improve Public—ICMA: https://icma.org/blog-posts/how-local-government-leveraging-technology-improve-public-engagement. Last accessed 30 Aug 2023
2. Riccio, E.L., et al.: Resultados do 13° contecsi usp congresso internacional de gest o da tecnologia e sistemas de informa. J. Inf. Syst. Technol. Manag. 13(2) (2015)

3. German, E.A.: Theoretical Innovation. Peter the Great St. Petersburg Polytechnic University, St. Petersburg (2018)
4. Bezrukikh, Yu.A.: Management Innovations as a Factor in the Introduction of New Technologies. Siberian State University named after. M. F. Reshetneva, Krasnoyarsk (2020)
5. World Bank Quality of public administration in the countries of the world. https://gtmarket.ru/research/worldwide-governance-indicators. Last accessed 30 Aug 2023
6. Ivanova, M., Kulkaev, G., Tanina, A.: Improving the UN methodology of the e-government development index. In: Ilin, I., Petrova, M.M., Kudryavtseva, T. (eds.) Digital Transformation on Manufacturing, Infrastructure and Service. DTMIS 2022. Lecture Notes in Networks and Systems, vol. 684, pp. 111–129. Springer, Cham (2023)
7. Bedenkova, A.S.: Digital Government as a Conceptual Development of Electronic Government. Polylogue, Moscow (2021)
8. Kulkaev, G.A., Mozaleva, N.I., Leontiev, D.N.: Analysis of Domestic and Foreign Experience of Project Management in the Implementation of the e-Government Concept, pp. 7–23. Public administration. Electronic Bulletin, Moscow (2022)
9. Kondratenko, V., Okopnyk, O., Ziganto, L., Kwilinski, A.: Innovation Development of Public Administration: Management and Legislation Features.Marketing and Management of Innovations, pp. 87–94 (2020)
10. Andrisani, P., Hakim, S., Savas, E.S.: The New Public Management: Lessons from Innovating Governors and Mayors. Springer, New York, NY (2002)
11. Indahsari, C., Raharja, S.: New Public Management (NPM) as an Effort in Governance. Jurnal Manajemen Pelayanan Publik 3(2), 73 (2020)
12. Rocha, J., Zavale, G.: Innovation and change in public administration. Open J. Soc. Sci. 9, 285–297 (2021)
13. Rocha, J., Araujo, J.: Administrative reform in Portugal: problems and prospects. Int. Rev. Adm. Sci 73, 583–596 (2007)
14. Madan, R., Ashok, M.: AI adoption and diffusion in public administration: a systematic literature review and future research agenda. Gov. Inf. Q. 40(4, 1) (2023)
15. Sudrajat, A., Andhika, L.: Empirical evidence governance innovation in public service. Jurnal Bina Praja 13, 407–417 (2021)

The Influence of the Social Environment on the Development of the Labor Market in the Field of Information and Communication Technologies (ICT)

Dmitriy Rodionov🆔, Irina Smirnova, Darya Kryzhko🆔, Olga Konnikova🆔, and Evgenii Konnikov🆔

Abstract Today's labor market, in line with the pace of life of the society, is changeable, oriented towards economic, social and technical development, open to qualitative modernization, ready to implement new ideas and tools. Reorganizations of requirements, trends and methods in labor market sectors related to data analysis and processing are especially swift. Specialists in the field of information and communication technologies are in demand. In today's realities, there are many accelerated targeted professional retraining programs—the resulting beneficial effect of which depends to a large extent on the student's knowledge base—which speaks of the high, underestimated in time, pace of expansion of areas of need for analysis at a new level. The interest in covering the growing demand—which is necessary, including for quality development—can be satisfied through the artificial regulation of key levers of society's influence on the number of ICT specialists. In the research, a mathematical description of the social environment along with statistical factors was formed and an analysis of the sensitivity of the ICT labor market to changes in a group of indicators that reflect public conditions was conducted. A model has been developed that allows for an assessment of the ICT job market, based on the reversibility and tempo of changing indicators of the social environment. The model identifies the parameters of the social environment, the artificial influence of which makes it possible to regulate the number of specialists in the field of information and communication technologies.

Keywords Labor market · Data analysis and processing · Information and communication technologies · Professional retraining programs

D. Rodionov · I. Smirnova · D. Kryzhko (✉) · E. Konnikov
Peter the Great St. Petersburg Polytechnic University, Polytechnicheskaya, 29, 195251 St. Petersburg, Russia
e-mail: darya.kryz@yandex.ru

O. Konnikova
St. Petersburg State University of Economics, Griboedov Canal 30-32, 191023 St. Petersburg, Russia

© The Author(s), under exclusive license to Springer Nature Switzerland AG 2023
Z. Dvořáková and A. Kulachinskaya (eds.), *Digital Transformation: What is the Impact on Workers Today?*, Lecture Notes in Networks and Systems 827,
https://doi.org/10.1007/978-3-031-47694-5_13

167

1   Introduction

In the conditions of an increased pace of social, technical and scientific development of society, the question of satisfying the changing—both from the point of view of structures and from the point of view of volumes and acceptance of new professions—the labor market is particularly acute. Modern technological trends require a confident response—from the state, enterprise, personality. The demand for the introduction of ICT into an ever greater number of areas of society life is growing—accordingly, the demand for qualitatively new technological solutions—the demand for qualified ICT staff—the demand for modern, fundamental, satisfying the changing infrastructure education—the demand for people able to get it, use it, develop it—the demand for time to get, apply and develop.

Despite the sharp contrast between the dynamism of the latest technologies and the strict conservatism of typical vectors of renovation and exchangeability of development trends—including trends in science and the labor market—the widespread introduction of advanced technologies into life today is the main step of progress, which is opposed only by a halt followed by stagnation [1–4].

In addition to—to some extent, frightening the public—the dynamism of technological changes is the problem of scientific gaps in research, assessment and description of the impact of I4.0 development on the activities of organizations—from small firms to state—there are a number of insufficiently analyzed risks accompanying the modernization of production and economic cycles [5].

These risks are associated with both an increase in unemployment, accompanied by an increase in demand in the labor market—when both supply and demand are excessive, but they do not correspond to each other, therefore, cannot be satisfied [6]; and with the typical peculiarities of new spheres of life—such as a geometric progression of cyberattacks, cyber security problems, and a lack of moral preparation of society sectors for industrial change [7].

Today's research focuses on the impact of ICT on the economy as a whole [8–10], on the environment [11], on CO_2 emissions [12, 13], on labor productivity [14], which demonstrates an understanding of the importance of information and communication technologies in modern reality. However, when looking at ICT as an instrument, a catalyst for some important social, economic and environmental indicators—for example, investments in ICT increase GDP [9, 15]—the statistics of describing the influence of various factors, which to some extent constitute the environment that ICT interacts with, on ICT itself is extremely incomplete.

Meanwhile, the ICT labor market is facing a number of serious structural problems, such as a lack of educational programs [16], inadequate educational services, a lack of timely responses to training problems [1]; a shortage of qualified workers [17], inequality in wages—lacking in a system that matches skills and their real value [18, 19].

At the same time, the social milieu has an immediate—and can't be denied—effect on the number of people employed in the ICT industry. In response to the ever-growing demand for competent ICT personnel, it is essential to understand which regulatory levers of the high-tech offering can be stimulated to increase the number of professional employees that meet the demand in the labor market, and which reforms may directly or indirectly lead to a decline in the employees of the information and communication technology sector [20].

Thus, the aim of this research is to provide a mathematical representation of a number of parameters describing the fraction of the social environment that hypothetically may influence the number of people employed in the ICT sector; testing assumed relationships between the social environment and information and communication technologies; and an aggregated picture of the obtained results through a mathematical model. The intention is to form an instrument allowing, within the social constraints—and with unspecified choose-and-pick factors—to regulate the number of ICT specialists, thus ultimately enabling the stimulation of technological, economic, and social progress.

2   Literature Review

At present, the IT technology sphere is a rapidly developing and attractive area. Many scholars are conducting research in the ICT field, assessing and analyzing its past and predicting its future. IT technologies are significant in terms of business strategies and financial forecasts for both individual organizations and the country, its economic growth and development strategies as a whole [8].

The world is on the verge of the Fourth Industrial Revolution [21]. Technological transformations are having a substantial impact on all spheres of society, including the economic structures at all levels. Researchers come to the conclusion that without the introduction of modern trends, tools, and methods in the field of IT, it is impossible to remain competitive in the social arena—both for medium-sized enterprises that address short-term budgeting and medium-term planning issues, as well as for advanced countries in terms of technological development, who need to form long-term strategies for progress, strengthen their positions in the international environment and protect national interests [8, 22, 23].

Investing in the ICT sector stimulates GDP growth—that is, an increase in the size of a country's economy. Through the prism of investing in Information and Communications Technology, economic growth of countries is considered [9, 10]. At the same time, the more developed the economy of the country, the stronger the effect of ICT investments on gross domestic product—for example, in South Korea, when the investment flow in information and communication technology is expanded by 1 percentage point of the country's GDP, it increases by 0.4% [15]. This study also asserts that the recent stagnation of ICT investments in South Korea is an indication of an economic downturn. This is evidence of the epochal importance of Information and Communications Technology in the current technological and social climate.

Speaking of the current socio-technological state, one cannot help but address the issue of COVID-19. The coronavirus infection has had a significant impact on society—including its social and economic spheres, the labor market, and health-care funding. On one hand, the pandemic has undeniably had a negative effect on many aspects of life worldwide. However, from the point of view of information and communication technology, COVID-19 has to some extent become a push for the improvement, implementation, and development of a range of significant contemporary and new technologies. Thus, in the context of the coronavirus infection and the distance work format of many market sectors, the technical skills of everyday employees have been in high demand. In response to this demand, the funding of ICT education as part of businesses' overall expenses has been forcibly increased—as the holding of training events aimed at stimulating the growth of their employees' information and communication competencies, accelerates the influence of their basic resources on the effectiveness of learning [24].

In the context of a COVID study, it was revealed that there was a positive effect of ICTs on the mental health of elderly people in pandemic conditions—households equipped with certain significant levels of Information and Communication Technologies (ICTs) have shown better psychological resilience to the difficulties caused by the spread of the coronavirus infection [25].

In addition, it should be noted that in relation to the ICT sphere, research related to mitigating the negative impact of the pandemic on economic wellbeing in enterprises, more advanced in terms of technological and software equipment, should be closed. Thus, in U.S. regions with a higher level of IT implementation among enterprises, a lower increase in unemployment in 2020 was observed [26]. An increase in the overall technological potential contributes to the increase in the immunity of enterprises to various external shocks, including those caused by COVID-19—such a strategy is most effective for firms that use more intangible capital [27]. And, in general, a more developed digital infrastructure leads to an increase in company revenues during crises [28].

Such a cushion of security in terms of risk in a relatively stable situation is an impetus for improvement and development. The introduction of the latest technologies can fundamentally change the activities of entire industries—achieving a certain level of technological potential allows companies to cope with the unpredictability of markets, simplify the complexity of a number of business processes, reduce the duration of innovation cycles. Companies gain control over supply chains, production processes, and weak links [1, 29, 30].

Due to the increased volume of information that requires quality processing, detailed analysis, and aggregated conclusions that can be implemented in strategies, the ICT professional market is expanding towards big data processing technologies. Currently, Big Data plays an important role, giving an immense advantage to companies working with it. Industry 4.0 aggregates the physical and digital spheres, combining traditional technologies, big data analytics, and modern digital technologies [31]. The introduction of big data analytics technologies is beneficial for companies since it increases their productivity, organizational efficiency, and accuracy of forecasting [32]. The development of BDA and other innovative technologies facilitates the work of businesses with a huge volume of data, allowing businesses to better use data and thus better understand their customers [33].

Today, the lack of understanding and limited resources of information analysis is the main hindrance to business development [2]. In addition, each technology sector requires a certain level of ICT, the influence of certain factors—and ultimately, the result due to the existing technological-organizational base is significantly different for businesses [34]. Business risks include difficulties in configuring advanced processes and digital technologies needed to create an intelligent working environment and transition to I4.0 [35]. In addition, numerous social and environmental risks are associated with the implementation of I4.0. These include resistance to learning the implementation of new technologies, ethical issues and security issues associated with the replacement of workstations with machines, as well as the fear of implementing intelligent systems throughout the entire value chain [5].

The abundance of different risks, as well as the periodic lack of clarity the need to implement ICT and difficulty to reorganize the management processes recently has led to the trend of overshooting the real demand for IT professionals than the offer. And often, there is a difference in the quality of the offer and the capability of the demand.

Technical progress, globalization, and the reorganization of the production process using outsourcing and offshore radically changed the demand for certain skills [19]. On the one hand, some jobs are disappearing while new ones are appearing; some of them are merely variations of existing jobs, while others are truly new jobs which didn't exist just a few years ago [17].

So, the labor market is challenged by a mismatched supply and demand for skills (For example, labor market research in Italy suggests that models described by career mobility theory seem to act only for over-educated and young workers with a higher education level, and particularly for males. Young workers with secondary education and female workers with higher education are at risk of falling into a trap of unemployment [16]. On the other hand, such figures are not linked to the reduction of jobs, but rather to an overall lack of—or even mismatch of—education of workers)—even though society is in the process of accepting the need for lifelong education, which is a response to not only the modernization of the global labor market, but also to the aging population in developed economies [3].

The quality of education is also significant from the point of view of economic strategies that rely on the use of low-skilled/low-paid labor as well as resource exports, which are not sustainable in the long term. To maintain the achieved development level and position on the international market, and even more so to increase these figures, new strategies are required, linked to the scientific and technological development and the qualitative reorganization of the infrastructure. It is precisely high-tech industries and R&D investments—which have been growing at a rapid rate since the beginning of the twenty-first century, sufficient for informed conclusions—that demonstrate these trends [36].

The important aspect of tech progress is the way it is managed—improving technological knowledge and science is most effective (and sometimes the only positive influence) when a well-structured management model is supported by quickly adjustable financing and flexible spending distribution mechanisms [37]. The real potential of advanced ICT lies in the exposure of the entire digital potential—i.e.

stepping beyond technology, as research [31, 38] confirms, creating the possibility of influencing society.

Thus, understanding the dependence of the ICT-sector, in terms of being populated with professionals, on the structure and stimuli of the social environment, is significant for the formation of efficient development strategies. Additionally, the adjustment of certain vectors of the social environment can serve as a tool for regulating the number of specialists in the ICT environment—that is, facilitate the creation of a favourable situation for technological growth in the country, be a method of stimulating progress in the ICT-sector specifically, and in the state as a whole.

3 Materials and Methods

Our research presents a model based on a centroid of people employed in the ICT sector. The factors influencing the centroid were identified based on the literature review presented. All of them are presented in a table, with their sources and their own symbolic representations (Table 1).

It is assumed that the factors will have the following effects on the centroid:

As internet users grow, the number of people employed in the ICT sector will increase, as this indicator can serve as an indicator of the information environment's interest in society—that is, it can reflect the population's quantitative predisposition to the development of information and communication technologies. Furthermore, due to the adoption of modern technological and digital trends worldwide, the number of internet users may be linearly related to a country's population (and migration, both in terms of scale and quality, as well as in terms of numbers, at present).

The level of education can describe the proportion of professionals in the ICT sector, as the ICT industry is highly technological and requires a quality preparation of the specialist. Accordingly, spending on education is a mediator of educational growth.

Similarly, the COVID-19 pandemic will have a positive effect: due to the specifics of transferring resources to the remote work format, the need for technical specialists capable of servicing such a transition is increasing; in addition, the ICT sector is inclined towards an online format, therefore there should not be a negative effect on the sector as a whole; and due to the reduction of the entire range of jobs (temporarily for the period of lockdown, or with long-term consequences) active population layers, facing the problem of employment, in search of earnings take the most in demand in the current realities and the most adaptable to the current restrictions niche—ICT specialists.

The impact of the coronavirus as a stimulus to the development of ICT can be reduced by a high level of health-care financing, provided that it is done correctly.

In turn, funding for healthcare is related to indicators of the social environment most clearly associated with life satisfaction. In the case of positive spending on health care, the rise in consumer prices, on the other hand, tends to be classified as a number of negative factors—an increase in living costs quantitatively, but not

Table 1 Summary table of indicators

Indicator	Notation	Unit of measure	Source
Climate related economic losses	X1	Million euro	Eurostat [39]
GDP per capita	X2	Million euro/population (thousand people)	Eurostat [39]
Life satisfaction	X3	Rating (0–10)	Eurostat [39]
CPI (consumer price index)	X4	Index (2015 = 100)	Eurostat [39]
House price index	X5	Index (2015 = 100)	Eurostat [39]
Health expenditure per capita	X6	Million euro/population (thousand people)	Eurostat [39]
Education expenditure per capita	X7	Million euro/population (thousand people)	Eurostat [39]
Level of education	X8	% of population	Eurostat [39]
Net migration rate	X9	Thousand people	Eurostat [39]
Population	X10	Thousand people	Eurostat [39]
Covid-19 Impact on ICT usage by size class of enterprise	X11	% of enterprises	Eurostat [39]
Individuals—internet use	X12	% of population	Eurostat [39]
Employed ICT specialists—total	Y1	Thousand people	Eurostat [39]

qualitatively, can obviously lead to dissatisfaction in society, thus reducing the real level of life satisfaction.

A similar impact can be attributed to the housing price index—however, this indicator is more global and reflects several other aspects of the social environment—if the consumer price index is offset by a rise in wages, inflation is justified, then the rapid rise in the housing price index (delayed due to the specifics of production—storage and procurement terms, payment algorithms for current projects) in the absence of obvious justifications and the impossibility of compensation causes a sharper dissatisfaction, having a greater negative effect on the degree of life satisfaction.

Changes in both the housing price index and the consumer price index can be partially attributed to the Gross Domestic Product (GDP), which in turn will have an obvious negative effect due to increasing climate-related costs.

Most of the indicators that describe the social environment can be linked to the level of life satisfaction (although this is not always convenient due to the rating

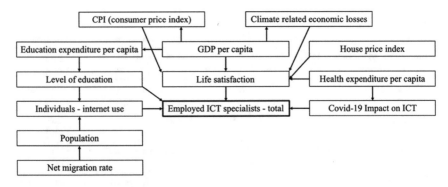

Fig. 1 Hypothetical conceptual model

format, subjective evaluations, and the presence of mistrust of some polling methods). Life satisfaction, as a guarantee of satisfaction of most basic needs, requires the satisfaction of increasingly complex social tasks, increasing the need for ICT professionals, increasing the interest in the ICT sector as a whole, and creating opportunities for individual development in a modern, technologically advanced environment, which can be reflected in the improvement of ICT skills.

The hypotheses presented are summarized in the conceptual model (Fig. 1).

The research is limited in sample size—31 European countries; and lack of differentiation of sources—all data is aggregated based on Eurostat statistics.

4 Results

A mathematical model describing people working in the ICT (y_1) field was derived from the analysis of the centroid study as follows:

$$y_1 = 53,7906 - 2,6982x_{11} - 3,5713x_{12} + 11,6975x_7 + 49,1384x_3,$$

where

x_{11}—Covid-19 Impact on ICT usage by size class of enterprise

x_{12}—Individuals—internet use

x_7—Education expenditure per capita

x_3—Life satisfaction. The presented factors explain 97.5% of the indicator.

The negative impact of COVID-19 on the number of people employed in the ICT sector has been confirmed, indicating the global seriousness of the pandemic's effects on the world community, even when the technology sector—which became one of the main paths for stabilizing the situation during the pandemic—had to bear a portion of the negative effect.

As the number of Internet users increases, the number of ICT professionals employed decreases, which may suggest the current dominance of low-level classification workers in the field of information and communication technologies, whose need is decreasing as new technological functionality is integrated—in response to the widespread development of infrastructure—and expanded into everyday life across humanity.

Investments in education—stimulating the structuring of the educational system according to the needs of the modern market, improving access to quality education, and expanding educational opportunities—directly addressing one of the most acute problems in the field of information and communication technologies—have a direct and positive impact on the centroid.

The number of people employed in the ICT sector increases with the level of life satisfaction. However, further research into this area needs to investigate the potential for a two-way influence between the ICT sector and the social environment. Life satisfaction as a highly subjective and sensitive indicator of the social environment may reflect the interplay between life satisfaction and ICT, thus laying the groundwork for a qualitative clustering of the countries researched in terms of levels of satisfaction—which, in the current sample, is not indicative of international social differentiation, as the study is confined to European countries many of whom are part of the European Union, whereby certain social indicators of these countries do not provide a vivid illustration of the international social differentiation of conditions.

Part of the 'Life satisfaction' factor, at 50.5%, was described by the model:

$$x_3 = 7,7762 + 0,0068x_4 - 0,0067x_5 + 0,8372x_6, \text{ where}$$

x_4—CPI (consumer price index)
x_5—House price index
x_6—Health expenditure per capita.

Increases in the consumer price index have a weak positive effect on life satisfaction. This is in contradiction with the hypothesis of a negative effect of growth in the consumer price index. Discussion of this issue is found in the respective section. It is noteworthy that this result can be most evidently influenced by the limitation of the sample size to only European countries, since most of them are at a comparable, not underestimated, level of development. Thus, a slight increase in the consumer price index may cause the reactions described in the discussion. For countries with lesser economic stability, lower levels of life quality and development, or a sharp increase in the consumer price index, the influence may exceed the ratio stated by the model.

The increase in housing price index has a negative effect on life satisfaction, indicating the need of the population of the sample countries to purchase real estate and the importance of this indicator to their mental state.

The increase in health care investments leads to an increase in life satisfaction, being the most significant factor of the model, which expresses the social nature of the satisfaction indicator, the strong dependence of it on the general well-being of the population, represented by the individual welfare of every particular inhabitant of the

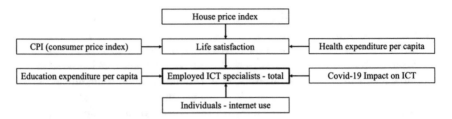

Fig. 2 The confirmed conceptual model

country. Thus, in future researches it is necessary to consider a similar social vector of influence factors for a qualitatively correct description of the research centroid.

The research study found that the effect of certain factors was not confirmed. The following optimized conceptual model was obtained (Fig. 2).

5 Discussion

The results of the study revealed the influence on the centroid of various social environment indicators such as the number of internet users, investments in education, the degree of impact of COVID-19, and life satisfaction, with the latter indicating the indirect influence of the consumer price index, housing price index, and healthcare investments on the centroid. Taken together, these factors described the resulting model index to 97.5%.

Increasing the proportion of enterprises affected by COVID-19 has a negative effect on the number of people employed in the ICT sector, which corresponds to key trends in the labor market since the beginning of the pandemic. The hypothesis that the specificity of information and communication technologies could mitigate the negative effect of coronavirus infection was not confirmed. However, it cannot be excluded that, at the general level, the singularity of the ICT sector contributed to a reduction of the effect.

The actual impact of the factor x_{12} (Internet use) doesn't match the hypotheses that have been made earlier either. An increase in the proportion of people using the Internet leads to a reduction in the number of people employed in the ICT sector. Such a result may indicate that due to the widespread use of basic technological resources, the need for technical specialists with basic ICT skills is decreasing in the labor market—the functionality of such specialists is distributed among employees, and familiarity with ICT is becoming an essential requirement in response to the current realities.

Education expenditures per capita, on the other hand, have a strongly positive effect on the centroid. Investments in population education contribute to an increase in the number of highly qualified technical specialists whose functions cannot be taken on by regular employees, thus leading to an increase in employment in the ICT sector.

The last factor in this model is the satisfaction with life. This is an aggregated indicator of the social environment in the country. Reflecting the general perception of the population of their own level of life, it serves as an indicator of the favourable environment for life in its most superficial sense—which includes, in detail, the predisposition to development, the state of all types of resources, and the ability to create new. Together, these vectors of society's conjuncture express an impulse potential that has developed in the country. Thus, the higher such potential, the greater the number of qualified specialists (in both "demand" and "supply" terms) on the labour market- that is, including more employed in the ICT sector.

According to empirical assumptions, the positive influence factors have a positive effect from the point of view of the typical subjective evaluation.

Life satisfaction increases with the consumer price index, forming an inverse hypothetic relationship. This influence can be explained by the fact that the price of the goods can be taken as one of the indicators of quality—that is, the higher the price, the better. Similar consumer policies have already been described in a number of studies. It was found that the price of the product is one of the significant parameters of the psychological definition of the quality of the product [40]. It is important from the point of view of compensatory thinking [41]—and despite the absence of a indicative correlation of price-quality in modern reality, consumers continue to take price as a positive factor in assessing the usefulness of the product [42]. Accordingly, with a slight increase of consumer price index within the financial possibilities of free coverage—economically stable consumers are psychologically satisfied with the acquisition of quality goods—accessing quality, and therefore their life satisfaction increases.

Decreasing life satisfaction with an increase in housing prices may be related to a decrease in the real availability (complicating the process) of purchasing housing ownership, which hinders the satisfaction of moral needs in their own homes and also prevents—from the consumer's point of view—one of the most psychologically comfortable ways of doing capital investment.

Increasing expenditure on health care—improving the quality of life and the degree of moral satisfaction of people through the level of medical care—contributes to an increase in life satisfaction.

6   Conclusion

The chapter examined the influence of the social environment, mathematically described by a set of statistical factors, on the labor market in the field of ICT, represented by the number of employees employed in the field of information and communication technologies.

A model was constructed that describes the character of the detected relationships. The model obtained allows for the assessment of the IKT potential, operating with known indicators of the social environment—life satisfaction, the level of investment in education, the number of Internet users and—in response to current market trends—the influence of COVID-19 on state enterprises. Also, life satisfaction directly was mathematically detailed through the following factors: consumer price index, housing price index and the level of investments in healthcare, which speaks of the indirect impact of these public sentiment indicators on the research centroid.

The formulated mathematical model allows, through investment in education and healthcare, control of housing prices and variability of life satisfaction, to adjust the number of ICT specialists within certain limits, thus modeling favourable conditions for perfecting technology and innovations - thus stimulating the country's qualitative economic growth.

The proposed research suggests to further describe the social environment using an extended range of indicators—to study the supply and demand correlation in the ICT job market, the influence of additional semi-professional courses aimed at improving skills in the ICT environment on the number of specialists. It is also recommended to pay attention to a number of external factors for the labor market of the state—such as the advantages of employment offers in the ICT sector in countries where the migratory flow of citizens of the state has been increased. It is possible that, from a certain level of national development, for the qualitative improvement of domestic specialists in the ICT spectrum, it is not enough to have only a general acceptable indicator of life satisfaction, but also a qualitative growth of the industry as a whole and the working conditions in it, in particular.

Acknowledgements The research is financed as part of the project "Development of a methodology for instrumental base formation for analysis and modeling of the spatial socio-economic development of systems based on internal reserves in the context of digitalization" (FSEG-2023-0008).

References

1. Ceipek, R., et al.: A motivation and ability perspective on engagement in emerging digital technologies: the case of Internet of Things solutions. Long Range Plan. **54**(5), 101991 (2021)
2. Lutfi, A., et al.: Drivers and impact of big data analytic adoption in the retail industry: a quantitative investigation applying structural equation modeling. J. Retail. Consum. Serv. **70**, 103129 (2023)
3. Hoffmann, E.: International statistical comparisons of occupational and social structures: problems, possibilities and the role of ISCO-88. In: Advances in Cross-National Comparison: A European Working Book for Demographic and Socio-Economic Variables, pp. 137–158). Springer US, Boston, MA (2003)
4. Rodionov, D., et al.: Methodology for assessing the digital image of an enterprise with its industry specifics. Algorithms **15**(6), 177 (2022)
5. Snieška, V., et al.: Technical, information and innovation risks of industry 4.0 in small and medium-sized enterprises–case of Slovakia and Poland. J. Bus. Econ. Manag. **21**(5), 1269–1284 (2020)
6. Horváth, D., Szabó, R.Z.: Driving forces and barriers of Industry 4.0: do multinational and small and medium-sized companies have equal opportunities?. Technol. Forecast. Soc. Chang. **146**, 119–132 (2019)

7. Birkel, H.S., et al.: Development of a risk framework for Industry 4.0 in the context of sustainability for established manufacturers. Sustainability **11**(2), 384 (2019)
8. Matyushok, V.M., Krasavina, V.A.: Global market of brand new information technology and the national interests. Natsional'nye interesy: prioritety i bezopasnost'= Natl. Interes.: Priorities Secur. **13**(11), 1988–2004 (2017)
9. Farhadi, M., Ismail, R., Fooladi, M.: Information and communication technology use and economic growth. PLoS ONE **7**(11), e48903 (2012)
10. Kwilinski, A.: Mechanism of modernization of industrial sphere of industrial enterprise in accordance with requirements of the information economy (2018)
11. Ollo-López, A., Aramendía-Muneta, M.E.: ICT impact on competitiveness, innovation and environment. Telematics Inform. **29**(2), 204–210 (2012)
12. Usman, A., Ozturk, I., Ullah, S., Hassan, A.: Does ICT have symmetric or asymmetric effects on CO2 emissions? Evidence from selected Asian economies. Technol. Soc. **67**, 101692 (2021)
13. Briglauer, W., et al.: Evaluating the effects of ICT core elements on CO2 emissions: recent evidence from OECD countries. Telecommun. Policy 102581 (2023)
14. Laddha, Y., et al.: Impact of information communication technology on labor productivity: a panel and cross-sectional analysis. Technol. Soc. **68**, 101878 (2022)
15. Sawng, Y.W., et al.: ICT investment and GDP growth: causality analysis for the case of Korea. Telecommun. Policy **45**(7), 102157 (2021)
16. Esposito, P., Scicchitano, S.: Educational mismatch and labour market transitions in Italy: Is there an unemployment trap? Struct. Chang. Econ. Dyn. **61**, 138–155 (2022)
17. Colombo, E., Mercorio, F., Mezzanzanica, M.: AI meets labor market: exploring the link between automation and skills. Inf. Econ. Policy **47**, 27–37 (2019)
18. Rodionov, D., et al.: Analyzing the systemic impact of information technology development dynamics on labor market transformation. Int. J. Technol. **13**(7) (2022)
19. Card, D., DiNardo, J.E.: Skill-biased technological change and rising wage inequality: some problems and puzzles. J. Law Econ. **20**(4), 733–783 (2002)
20. Rodionov, D., et al.: The information environment cluster distribution of the regional socio-economic systems in transition economy. In: International Scientific Conference "Digital Transformation on Manufacturing, Infrastructure and Service", pp. 203–217. Springer Nature Switzerland, Cham (2022)
21. Tamvada, J.P., et al.: Adopting new technology is a distant dream? The risks of implementing Industry 4.0 in emerging economy SMEs. Technol. Forecast. Soc. Chang. **185**, 122088 (2022)
22. Rodionov, D., et al.: Impact of COVID-19 on the Russian labor market: comparative analysis of the physical and informational spread of the coronavirus. Economies **10**(6), 136 (2022)
23. Yuan, S., Pan, X.: The effects of digital technology application and supply chain management on corporate circular economy: a dynamic capability view. J. Environ. Manage. **341**, 118082 (2023)
24. Karaca, A., Aydogmus, M.E., Gunbas, N.: Enforced remote work during COVID-19 and the importance of technological competency: a job demands-resources perspective. Eur. Rev. Appl. Psychol. 100867 (2022)
25. Nedeljko, A.M.: The use of information and communication technologies affects mental health and quality of life of older adults during the COVID-19 pandemic. Ifac-papersonline **55**(10), 940–945 (2022)
26. Pierri, N., Timmer, Y.: IT shields: Technology adoption and economic resilience during the COVID-19 pandemic. Available at SSRN 3721520 (2020)
27. Ai, H., et al.: News shocks and the production-based term structure of equity returns. Rev. Financ. Stud. **31**(7), 2423–2467 (2018)
28. Doerr, S., et al.: Technological capacity and firms' recovery from Covid-19. Econ. Lett. **209**, 110102 (2021)
29. Rodionov, D.G., et al.: Information environment quantifiers as investment analysis basis. Economies **10**(10), 232 (2022)
30. Fareri, S., et al.: Estimating Industry 4.0 impact on job profiles and skills using text mining. Computers in industry **118**, 103222 (2020)

31. Liao, Y., et al.: Past, present and future of Industry 4.0-a systematic literature review and research agenda proposal. Int. J. Prod. Res. **55**(12), 3609–3629 (2017)
32. Dubey, R., et al.: Big data analytics and artificial intelligence pathway to operational performance under the effects of entrepreneurial orientation and environmental dynamism: a study of manufacturing organisations. Int. J. Prod. Econ. **226**, 107599 (2020)
33. Youssef, M.A.E.A., Eid, R., Agag, G.: Cross-national differences in big data analytics adoption in the retail industry. J. Retail. Consum. Serv. **64**, 102827 (2022)
34. Wang, Y.N., Jin, L., Mao, H.: Farmer cooperatives' intention to adopt agricultural information technology—mediating effects of attitude. Inf. Syst. Front. **21**, 565–580 (2019)
35. Lee, J., Bagheri, B., Jin, C.: Introduction to cyber manufacturing. Manuf. Lett. **8**, 11–15 (2016)
36. Lall, S., Hamdi, M.: Investment and technology policies for competitiveness: Review of successful country experiences. UN (2003)
37. Peña-López. I. et al.: OECD digital economy outlook (2015)
38. Tseng, M.L., et al.: Circular economy meets industry 4.0: can big data drive industrial symbiosis?. Resour. Conserv. Recycl. **131**, 146–147 (2018)
39. Eurostat, https://ec.europa.eu/eurostat/web/main/data/statistical-themes. Last accessed 25 May 2023
40. Dodds, W.B., Monroe, K.B., Grewal, D.: Effects of price, brand, and store information on buyers' product evaluations. J. Mark. Res. **28**(3), 307–319 (1991)
41. Scitovszky, T.: Some consequences of the habit of judging quality by price. Rev. Econ. Stud. **12**(2), 100–105 (1944)
42. Völckner, F., Hofmann, J.: The price-perceived quality relationship: a meta-analytic review and assessment of its determinants. Mark. Lett. **18**, 181–196 (2007)

Impact of the External Environment on the Development of the ICT Labor Market

Dmitriy Rodionov⬤, Darya Kryzhko⬤, Irina Smirnova, Olga Konnikova⬤, and Evgenii Konnikov⬤

Abstract Today, the ICT market is developing rapidly. At the same time, a number of studies have proved that the human potential in this market does not always keep pace with the development of the technological component. The aim of the current study is to identify what environmental external factors have an impact on the development of the ICT market. The study was based on open data for 22 developed countries with a mature ICT market. 14 factors were included in the conceptual model. By building regression models on the available statistical data, it was proved that share of ICT specialists in the total number of employees is influenced by such factors as the Number of ICT companies, English language Proficiency Index, Public use of the Internet indicator and Brain Drain Index. At the same time, Number of ICT companies itself is influenced by Index of economic freedom, Index of individualism and amount of Venture investments. These findings highlight the interconnectedness of different factors within the ICT market ecosystem. Factors such as the number of ICT companies, economic freedom, language proficiency, and public usage of the Internet play crucial roles in shaping the employment landscape for ICT specialists. Additionally, the level of individualism and investment activity also impact the growth and development of the ICT market. By understanding and analyzing these environmental factors, policymakers, stakeholders, and industry professionals can make informed decisions to foster the growth and development of the ICT sector.

Keywords ICT labor market · Human potential · ICT market ecosystem · External factors · ICT specialists · Regression models

D. Rodionov · D. Kryzhko (✉) · I. Smirnova · E. Konnikov
Peter the Great St.Petersburg Polytechnic University, Polytechnicheskaya, 29, 195251 St.Petersburg, Russia
e-mail: darya.kryz@yandex.ru

O. Konnikova
St. Petersburg State University of Economics, Griboedov Canal Emb., 30-32, 191023 St. Petersburg, Russia

© The Author(s), under exclusive license to Springer Nature Switzerland AG 2023
Z. Dvořáková and A. Kulachinskaya (eds.), *Digital Transformation: What is the Impact on Workers Today?*, Lecture Notes in Networks and Systems 827,
https://doi.org/10.1007/978-3-031-47694-5_14

1 Introduction

The development of ICT is a global trend of scientific and technological progress in recent decades. As an engine of economic change for more than two decades, the ICT sector is a key determinant of a country's competitiveness, attracting investment and creating innovation. The ICT sector plays a strategic role in promoting growth and competitiveness in the whole world, especially in the developed economies due to the processes of generating, exchanging, storing information, as well as creating various communications for the interaction of economic entities. ICTs today occupy a central place in the innovative development of key areas of the life of society: municipal and state administration, education, business, healthcare, security, culture and public life.

The impact of ICT industries is critical to improving productivity and efficiency. In economics, the spillover effect is a situation when some economically significant events lead to the emergence of others, while the two contexts under consideration may seem unrelated. The contribution of the ICT sector spillover effect to productivity growth in the whole eco is significant. Research results show that the spread of ICT plays an important role in increasing productivity in the EU countries. An increase in the country's ICT index by 1% will lead to an increase in labor productivity in the country by an average of 0.357%, an increase in labor productivity in all other countries by an average of 0.421% [1].

Thus, it can be argued that a mature ICT sector is needed in order to benefit from digitalization, keep up with competitors in globalized markets, and establish technology leadership. As a result, the economic importance of ICT especially for developed countries continues to grow.

The active financial development of the ICT sector is due to the growth of young ICT companies, the focus of employers on improving working conditions and the automation of routine duties. ICT sector is distinguished by good career prospects, interesting tasks, and specialists in this field are rapidly gaining popularity and demand.

At the same time, to ensure the stable development of the ICT sector, a certain balance in the labor market in the industry is required. Regulation of supply and demand for vacancies and services, formation of labor relations regulations, uniform development of existing and new areas of ICT. It is also important for employers to ensure the necessary technological equipment of the workplace, the availability of work for different categories of specialists [2].

At the same time, the labor market in the ICT sector is directly related to the state of the external economic environment in the region. For example, the unfavorable economic environment in the country is forcing companies operating in the ICT industry to focus on short-term reduction in operating costs, rather than on increasing revenue and efficiency, which leads to changes in the balance of supply and demand in the labor market and, as a result, slows down development of the ICT sector as a whole.

Taking into account all of the above, the main aim of this research is to explore the dependence of the labor market in the ICT sector on the external economic environment.

2   Literature Review

Any professional-branch segment of the labor market is under the influence of various political, economic and social factors. Of course, this is also true for the labor market of the ICT sector. At the same time, the factors influencing the sectoral segments of the labor market will differ depending on the characteristics of the functioning of the industry itself [3].

The ICT labor market supply will be characterized by the share of employed population in the industry. Particularly in the economies of developed countries, the ICT population represents a growing proportion of the total employed population. This growth is driven primarily by the growing demand for ICT professionals and is driven both by existing companies that are hiring more people to expand their business, and by new companies providing ICT services. At the same time, it is important to note that it is new companies that create the vast majority of jobs in the ICT labor market [4].

An important driver of this intense growth in the ICT sector is increased productivity in the digital economy. The total factor productivity in the ICT sector is greatly influenced by intangible assets [5], in particular, the company's human capital [6]. Research results show that human capital potential, human capital knowledge and human capital skills have a significant positive relationship with organizational performance [7].

Thus, it can be argued that in order to maintain the growing share of employment in the ICT sector in the future, a constant supply of appropriately qualified ICT specialists is required. It is important to note that recent studies of supply and demand in the labor market have shown that the ICT sector is experiencing disequilibrium growth—technologies in ICT are developing faster compared to the growth of human capital skills in the industry [8, 9].

It is worth taking into account the fact that the geographical location of individual labor markets is characterized by a certain set of factors that ultimately determine who can participate in the supply of labor on the market and under what conditions. Consequently, the labor markets of different countries differ in various parameters [10]. This is true even if we are talking only about developed countries.

In particular, English language proficiency can be an important driver of supply growth in the ICT labor market. English dominates programming languages, industry education programs and ICT research. Indeed, all the most used programming languages, including Java, C++, Python and JavaScript, are written in English, even though some of them, like Python, were not created in English-speaking countries [11].

Moreover, most of the online content is generally in English. This can be a major barrier to the dissemination of ICTs in different countries, as information content on the Internet and in software is not always available in local languages. The mismatch between the current language of the population and the language of ICT creates a cultural and technological mismatch that hinders the use of ICT. It can also be noted that a high level of English proficiency mainly characterizes well-educated segments of the population [12].

At the same time, the educational background of the founder or team members does not necessarily have a positive effect on productivity, which is significant for maintaining a growing share of those employed in the field of ICT, if this education is not supplemented by skills gained through experience [13]. This is confirmed by the results of a study of the importance of factors and their role in the success of high-tech start-ups, which showed that the attitude and ability of the core team are of paramount importance [14].

So, if we talk about the importance of personal qualities and abilities of specialists, then, in addition to a positive relationship with education, it was found that individualism also has a positive relationship with career adaptation. At the same time, career adaptation has been identified as one of the skills necessary for a person working in the field of ICT, due to the rapid and consistent technological modernization [15]. The construction of individualism-collectivism considered in the study represents the individual "I" and the types of groups to which, in the opinion of a particular individual, he belongs [16]. From this point of view, individualism can be seen as a catalyst for starting one's own business in the ICT sector.

In addition, the age of ICT professionals themselves and entrepreneurs working in the industry can play a certain role in the share of ICTs employed in the labor market. Research results suggest that the likelihood of creating a startup decreases as the age of the entrepreneur increases [17]. At the same time, an increase in the supply of young workers is positively associated with the creation of new firms in high-tech industries [18, 19].

If we talk about factors directly related to the technical characteristics of the labor market in the ICT sector, then the prevalence of Internet use by the population is important for maintaining the supply of labor in the ICT. The use of the Internet can also indirectly influence the ICT industries, as the population, through horizontal information exchange, becomes more aware of the ICT and can take a proactive part in the provision of services [20].

At the same time, the use of the Internet is also characterized by significant social and spatial digital divides, that is, inconsistencies in terms of the ease and cost of accessing the Internet at several spatial scales [1]. On a global scale, Internet access accurately reflects geographic uneven development and related disparities in wealth and power. However, the digital divide also exists within countries. Although the vast majority of people in economically developed countries have access to the Internet, especially in places like Scandinavia where Internet use is almost universal. Poor and rural areas still show lower rates of use [21].

The growth of labor supply in the field of ICT can be influenced to a certain extent by the average salary of ICT specialists. In the European Union, wage levels for skilled ICT professionals vary more between countries than within countries. At the same time, the leadership of the United States in the world market in the field of ICT has led both to a significant "brain drain" of highly qualified personnel in the field of ICT, and to the neglect of many proposals in the domestic markets of countries [22]. Thus, we can say that the "brain drain" also affects the supply of labor in the ICT sector.

The high demand for ICT professionals has led to strong competition for experienced people, although wage levels have not increased dramatically, except for specific skills such as specialized developers and security experts [23, 24]. This is due to the high concentration of candidates in the ICT labor market. Empirical evidence presented in studies on the impact of labor market concentration on wages suggests that wages are declining in more concentrated labor markets [25, 26].

At the same time, the direct impact of the level of wages in the industry on labor supply is ambiguous. The substitution effect is always positive—higher wages cause an increase in labor supply. But the income effect is always negative—higher wages imply higher income, and higher income implies greater demand for leisure, which in turn leads to a decrease in the amount of labor offered [27].

Moreover, an entrepreneurial ecosystem characterized by low levels of business and financial freedom cannot provide the financial support needed for fast-growing enterprises [28]. Higher taxes and discriminatory tariffs imposed by the state on ICT goods and services reduce the level of their adoption by both enterprises and consumers [29], and, consequently, the demand for ICT services in general. Studies also show that restrictions on technology exports reduce a firm's investment in R&D [30], which is important for high-tech companies in the ICT sector. Reducing import tariffs, on the contrary, significantly increases the productivity of ICT companies in the country by optimizing the structure of production factors, importing more high-quality inputs and increasing investment in research and development [31]. In addition, the level of economic freedom in a country affects the flexibility of the labor market as a whole. Research findings suggest that labor market flexibility, in turn, affects the inflow of foreign direct investment, which is often necessary for ICT companies to successfully enter the market and continue to exist [32].

However, there is also an inverse relationship: the Internet and mobile technologies increase the level of democracy and, as a result, business freedom. In particular, freedom of the press and the introduction of ICT complement each other in developed countries. The effectiveness of ICTs in promoting political and economic freedom increases as more users connect to ICT networks [33]. A well-functioning bureaucracy and a state that supports entrepreneurs have a positive impact on entrepreneurship [34].

A firm's investment decisions in the ICT industry are highly sensitive to the availability of cash flow. Previous research has empirically confirmed that the use of debt in high-tech firms is rare. This is due to a number of factors that increase the uncertainty associated with the provision of loans in the ICT sector [35]. With strong network effects in the ICT sector, which tend to increase the volatility of a firm's future cash flows, it is difficult to predict its success. This volatile cash flow pattern increases the cost of external funds and reduces the firm's ability to make new investments [36]. The nature of some ICT products is characterized by fixed entry costs and low marginal costs [37], which can make it difficult for all investors, including lenders, to recover fixed entry costs.

Due to the above facts, the main source of external financing for ICT companies is equity financing, such as, for example, venture capital, foreign direct investment, etc. [38]. It is the provision of large venture investments that is widely considered as a factor that distinguishes successful start-ups in the sector from unsuccessful ones [39, 40].

Venture investment and government support for entrepreneurs together contribute to the high growth potential of ICT start-ups by providing promising ICT enterprises with sufficient and long-term funding to cope with the delay between product or service development and market entry, as well as allowing entrepreneurs to start at a critical size from the point of employment perspective [41].

3 Methodology

All factors of external environment influencing the development of the labor market in the field of ICT, according to the results of the theoretical review, can be combined into the following groups: the wide-spread of professional education, the economic development of the country and the information development of the country. The constructed conceptual model (Fig. 1) analyzes the impact of groups of these factors on the ICT labor market. The sample is based on 22 developed countries (Australia, Austria, Belgium, Canada, Denmark, Finland, France, Germany, Ireland, Israel, Italy, Japan, Netherlands, New Zealand, Norway, Portugal, Singapore, Spain, Sweden, Switzerland, United Kingdom, United States) for 2022.

To build a conceptual model, the following hypotheses were formulated:

- Public and private investments in education and educational infrastructure create favorable conditions for obtaining high-quality knowledge in the field of ICT and, as a result, increase the number of qualified personnel.
- The rapid development of ICT technologies and the high demand for information services form a competitive salary, which contributes to the growth of specialists in the market.
- Globalization integrates many countries, brings people together in business and educational activities, English becomes the universal language, this creates great opportunities for increasing the number of ICT specialists.
- The spread of the Internet allows people of different ages to use the worldwide web, to learn new digital professions and to replenish the talent pool of ICT specialists.
- The ability to work remotely attracts IT professionals and allows to work from anywhere in the world.

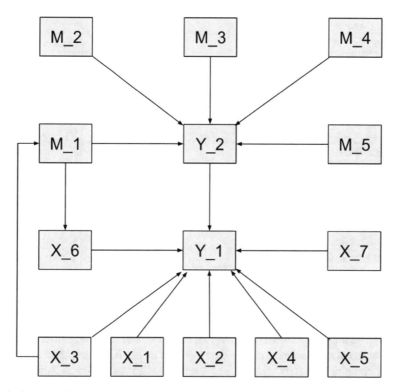

Fig. 1 Conceptual research model

- Difficult political situation in the country, low economic development leads to the emigration of the country's intellectual potential, thereby reducing the number of ICT specialists.
- The promise of the ICT sector attracts entrepreneurs to open new companies, creating new jobs for specialists.
- The income tax rate affects the amount of retained earnings and, as a result, the rate value allows to make a decision to open a business.
- The number of companies affects the volume of ICT market revenue.
- Individualism is the main component of entrepreneurship and, as a result, the catalyst for starting your own business.
- Economic freedom, namely the freedom of professional, consumer, property, financial choices, as well as the freedom of entrepreneurial activity, create favorable conditions for the establishment of new ICT companies.
- Venture investment increases the number of ICT companies.
- State support programs create favorable conditions for founding new ICT companies.
- The use of the Internet increases the economic freedom of ICT professionals by providing access to news and educational resources.

Table 1 Summary array of indicators

Indicator	Designation	Measure	Source
Share of ICT specialists in the total number of employees	Y_1	%	[42]
Number of ICT companies	Y_2	Units	[43]
Average salary in the ICT sector	X_1	$	[44]
English language Proficiency Index	X_2	1–800	[45]
Public use of the Internet	X_3	%	[46]
Education index	X_4	0–1	[47]
Global remote work index	X_5	0–1	[48]
Brain Drain Index	X_6	0–10	[49]
Average age	X_7	Years	[50]
Index of economic freedom	M_1	%	[51]
Corporate income tax rate	M_2	%	[52]
Index of individualism	M_3	1–100	[53]
Venture investments	M_4	mln $	[54]
Availability of ICT industry support programs	M_5	0–4	[55]

Based on these hypotheses, a conceptual model was built that describes the impact of the external environment on the development of the ICT labor market (Fig. 1). Table 1 contains summary array of the presented indicators with the identification of each data source.

4 Results

The proposed conceptual model was through using regression analysis in KNIME software.

As a result of the study, the following optimized model was obtained:

$$y_1 = -0,0429 - 5,99E(-5) * x_2 + 0,1382 * x_3 + 0,0069 * x_6 - 6,42E(-8) * y_2$$

$$y_2 = 254782,395 - 4525,24 * m_1 + 1,661 * m_3 + 1816,805 * m_4$$

$$m_1 = 25,855 + 54,886 * x_3$$

The coefficient of determination is 51.46%; the approximation error is 0.133. Spread of residuals relative to the actual values taken by the considered endogenous variable reflects the heteroscedasticity of the distribution (heterogeneity of observations). In accordance with the resulting elasticities, the greatest impact on the share of those employed in the ICT sector is the use of the Internet; the smallest is the number of ICT companies. The influence of the brain drains and the level of English proficiency is significant.

As for the model with "Number of ICT companies" resulting indicator, it looks the following way:

$$y_2 = 254782, 395 - 4525, 24 * m_1 + 1, 661 * m_3 + 1816, 805 * m_4$$

The proportion of explained variance is 88.67%. The approximation error is 0.624. Spread of residuals relative to the actual values taken by the considered endogenous variable shows the distribution is heteroscedastic. Analysis of the elasticity of variables shows that the individualism index and economic freedom have the greatest impact on the number of ICT companies; the smallest impact takes venture investments.

The influence of exogenous factors on the index of economic freedom describes the resulting equation:

$$m_1 = 25, 855 + 54, 886 * x_3$$

The coefficient of determination is 0.3167, the approximation error is 0.047. The spread of residuals relative to the actual values taken by the considered endogenous variable shows the heteroscedasticity of the dependence of the residuals. The lower limit of the elasticity coefficient is 0.44165, the upper limit is 0.87368.

Based on the results of checking the proposed conceptual research model based on the regression analysis method, the result optimized conceptual model was obtained (Fig. 2).

5 Discussion

Based on the analysis in the Research Results section, a validated conceptual model was formed. The model contains the following factors: economic freedom index, individualism index, venture investments, brain drain index, English language proficiency index, Internet use by the population, the share of ICT specialists in the total number of employees and the number of IT companies.

It follows from the proven hypotheses that the number of ICT companies in developed countries is related to the index of economic freedom. Indeed, the ICT sector is characterized by the development of various start-ups, and the representatives themselves are distinguished by individualization of work and independent decision-making, which is confirmed by the index of individualism.

In the field of ICT, advanced technologies are being created, there is an act growth of financial and technical indicators, these events increase investment attractiveness and motivate investors.

The influence of economic freedom on the number of active companies in the ICT market is also confirmed. This proves the results of previous studies showing

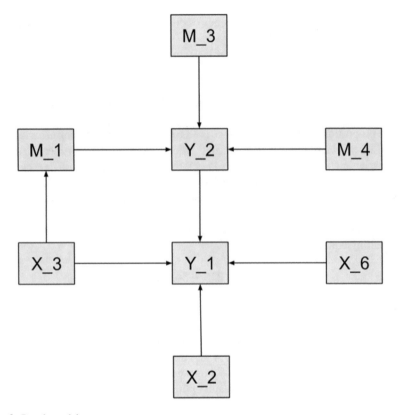

Fig. 2 Result model

that the absence of inflated taxes and tariffs on imports improves the productivity of ICT companies and the development of the ICT industry as a whole, and also helps attract critical investments for companies in the industry.

The number of ICT companies, which reflects the demand in the ICT labor market, in turn affects the supply of labor, expressed as the share of ICT specialists in the total number of employees.

The spread of the Internet is a truly important factor in shaping the supply of labor in the ICT market, since a large percentage of work is done online. Moreover, the widespread use of English allows employees to interact with colleagues from different parts of the country and the world. In the long term, integration can lead to the formation of new companies and the creation of new jobs in a rapidly growing industry, this is confirmed by the excess of the number of new companies over existing ones.

Otherwise, the unstable situation in the country can lead to the emigration of the country's intellectual capital, in other words, economic, political, socio-cultural inconveniences lead to a brain drain.

The impact of government support programs on the number of ICT companies and, consequently, on the demand in the ICT labor market has not been confirmed. This contradicts a number of studies cited in the literature review, which consider the effective functioning of state support programs as one of the factors in the growth in the number of new ICT companies, which make up the largest number of companies in the market. The industry accounts for a significant amount of money turnover and scientific developments, which ensures a certain autonomy of activity. We also note that ICT is at the development stage and is a relatively new industry, the business processes of which are not yet fully regulated. These facts create obstacles and require more time to develop a state support program.

According to the results, the tax rate directly influences on the company profits, in contrast to taxes and tariffs on imports included in the model as an integral part of the indicator of business freedom in the country, also does not affect the number of ICT companies operating in the market. This may be due to the fact that although the reduction in the tax rate entails an increase in free cash that a company can invest in business development, it does not always contribute to the creation of new companies in the market. Despite the reduction in the payback period of business for investors, which is important for gaining a foothold in the ICT sector and is facilitated by lower tax rates on corporate profits, they also contribute to the monopolization of the market by turning large market companies into even larger holdings and transnational corporations. As a result, for a novice entrepreneur, competition in the market is intensifying. Thus, tax incentives are not in all cases an incentive tool. Therefore, when starting an ICT business, an entrepreneur takes into account not so much the tax burden as competition, the growth of which can change the entrepreneur's decision to start his own business.

The impact of average wages on supply growth in the ICT labor market has also not been confirmed. This, in general, is consistent with the studies already presented in the field, and confirms that the simultaneous effects of substitution and income on the labor market in the ICT industry cancel each other out. At the same time, the indirect impact of differences in wages in various ICT sectors across countries on the outflow of specialists from the country is partially represented in the model by the "brain drain" factor.

The impact of the country's education index on ICT labor supply has not been confirmed. The model used the overall education index for the country. The absence of its influence means that the development of the ICT labor market is not directly related to the overall share of educated people. This confirms the conclusions of previous studies on the importance of factors and their role in the success of high-tech start-ups in the ICT industry that specialized education plays a minor role, if not backed by experience, and is also related to the prevailing opinion that it is better to get a job as an intern and to comprehend the basics of the profession in practice. The attitude of employers to the qualifications of potential employees can also play an important role. Many employers indicate that the priority when considering candidates for a position is the completion of a test task and the result obtained, and education is of secondary importance. It is also worth noting that the academic program of universities does not have time to take into account all the changes in the rapidly developing ICT industry, so the quality of higher education does not always meet the expectations of companies and, as a result, does not necessarily contribute to the growth of those employed in the industry.

The influence of the possibility of remote work on the number of specialists employed in the ICT industry has not been proven. The factor was calculated based on the global remote work index, which includes cybersecurity, economic and social conditions, digital and physical infrastructure. The lack of influence may be due to the need for technological equipment of the workplace, the impossibility of disposing of corporate information outside the office.

The influence of the average age of the population on the growth in the share of IT specialists employed in the economy has also not been confirmed. Note that the average age of the population is generally not necessarily related to the average age of specialists in the industry. Moreover, there is a trend of complementarity in the ICT industry: young specialists, on the one hand, establish breakthrough startups, make global scientific research and discoveries, fill the industry with energy and new goals, and more experienced specialists, on the other hand, focus on attracting investments, refining and the introduction of new ideas, as well as the formation of large companies that generate intellectual capital.

6 Conclusion

The aim of the research was to investigate the dependence of the labor market in the ICT industry on the external economic environment.

The share of those employed in the ICT sector is most affected by the use of the Internet; the smallest impact comes from the number of ICT companies. Brain drains and English language proficiency also have a significant effect. Consequently, the number of IT specialists directly depends on the use of the Internet, this is due to the specifics of the work, which can only function within its limits. A slight relationship with the number of companies may indicate the priority of individual work among programmers and developers. English skills and brain drain are inextricably linked with the possibility of relocation of specialists and the increase in the employed population in other countries. The individualism index and economic freedom have the greatest impact on the number of ICT companies; the smallest is venturing investments.

The result obtained indicates that the founders of companies are people with developed entrepreneurial abilities, distinguished by the presence and expression of their own point of view. The degree of economic freedom determines the confidence in making decisions, building financial plans, and also creates the basis for the implementation of ideas. At the same time, venture investments have little impact and this is due to the economic independence of the industry.

Thus, on the basis of the study, it follows that the labor market in the ICT industry really depends on the external economic environment.

The ICT labor market is dynamic and evolving, influenced by many factors. During the study, the main ones were identified: the index of economic freedom, the index of individualism, venture investments, the brain drain index, the English language proficiency index, the use of the Internet by the population, the share of ICT specialists in the total number of employees and the number of IT companies.

The obtained results explain a super-new reality of the ICT labor market, and allow to make a forecast about its further development trends:

- The economic and political situation will intensify the trend towards the relocation of individual specialists and entire companies, fluctuating the statistics of the number of employees and ICT companies by country.
- The development of economic freedom will attract new talented personnel to the ICT industry, who will make a breakthrough in the development of high technologies.
- Dynamics of open vacancies and posted resumes on job search sites will fluctuate, but tend to rise due to the globalization of the market and the widespread use of the English language, the desire of company leaders and professionals to create partnerships on a global level.
- Competition in the ICT market will increase, which will affect the number of ICT companies, but the possibility of venture capital investment will become a motivation for owners to create a unique product.
- The desire for individualism will lead to an increase in the number of entrepreneurs and self-employed, since this category is distinguished by financial and psychological independence.

Summing up the above, we note that the labor market of ICT sector has great potential, unites many people and makes it possible to realize bold ideas.

The limitation of the study is due to the fragmentation of data, the difference in methods for calculating economic indicators in different countries, as well as the availability of statistical data.

Acknowledgements The research is financed as part of the project "Development of a methodology for instrumental base formation for analysis and modeling of the spatial socio-economic development of systems based on internal reserves in the context of digitalization" (FSEG-2023-0008).

References

1. Goldfarb, A., Prince, J.: Internet adoption and usage patterns are different: implications for the digital divide. Inf. Econ. Policy **20**(1), 2–15 (2008)
2. Rodionov, D., et al.: Analyzing the systemic impact of information technology development dynamics on labor market transformation. Int. J. Technol. **13**(7) (2022)
3. Egorshin. A.P.: Motivation of Labor Activity. N. Novgorod, NIMB, 320 (2003) (in Russian)
4. Gabison, G.A.: Birth, survival, growth, and death of ICT companies (No. JRC94807). Joint Research Centre (Seville site) (2015)

5. Nakatani, R.: Total factor productivity enablers in the ICT industry: a cross-country firm-level analysis. Telecommun. Policy **45**(9), 102188 (2021)
6. Shestakova, I.G.: Human capital in the digital age. Sci. J. NRU ITMO Ser. Econ. Environ. Manage. **1**, 56–63 (2018) (in Russian)
7. Aman-Ullah, A., et al.: Human capital and organizational performance: a moderation study through innovative leadership. J. Innov. Knowl.Knowl. **7**(4), 100261 (2022)
8. Rodionov, D., et al.: Impact of COVID-19 on the Russian labor market: comparative analysis of the physical and informational spread of the coronavirus. Economies **10**(6), 136 (2022)
9. McLaughlin, S., et al.: E-skills and ICT professionalism. Fostering the ICT profession in Europe. Final Report (2012)
10. Carmody, P.: Labor market (2020)
11. Graves, A.: The linguistic dominance of English in computer science: a case study of Francophone programmers. French in the world (2019)
12. Grazzi, M., Vergara, S.: ICT in developing countries: are language barriers relevant? evidence from Paraguay. Inf. Econ. Policy **24**(2), 161–171 (2012)
13. Santisteban, J., Mauricio, D.: Systematic literature review of critical success factors of Information technology startups. Acad. Entrepr. J. **23**, 1–23 (2017)
14. Chorev, S., Anderson, A.R.: Success in Israeli high-tech start-ups, critical factors and process. Technovation **26**(2), 162–174 (2006)
15. Omar, S., Noordin, F.: Moderator influences on individualism-collectivism and career adaptability among ICT professionals in Malaysia. Proc. Econ. Financ. **37**, 529–537 (2016)
16. Hofstede, G.: Dimensionalizing cultures: The Hofstede model in context. Online Read. Psychol. Cult. **2**(1), 8 (2011)
17. Oakey, R.P.: Technical entreprenenurship in high technology small firms: some observations on the implications for management. Technovation **23**(8), 679–688 (2003)
18. Rodionov, D., et al.: Methodology for assessing the digital image of an enterprise with its industry specifics. Algorithms **15**(6), 177 (2022)
19. Ouimet, P., Zarutskie, R.: Who works for startups? the relation between firm age, employee age, and growth. J. Financ. Econ. **112**(3), 386–407 (2014)
20. Pascu, C., et al.: The potential disruptive impact of Internet2 based technologies. First Monday (2007)
21. Chakraborty, J., Bosman, M.: Measuring the digital divide in the United States: race, income, and personal computer ownership. Prof. Geogr.Geogr. **57**(3), 395–410 (2005)
22. Parthasarathi, A.: Tackling the brain drain from India's information and communication technology sector: the need for a new industrial, and science and technology strategy. Sci. Public Policy **29**(2), 129–136 (2002)
23. Rodionov, D., et al.: The information environment cluster distribution of the regional socio-economic systems in transition economy. In: International Scientific Conference "Digital Transformation on Manufacturing, Infrastructure & Service", pp. 203–217. Springer Nature Switzerland, Cham (2022)
24. Kolding, M., et al.: Information management–a skills gap? The Bottom Line **31**(3/4), 170–190 (2018)
25. Izumi, A., Kodama, N., Kwon, H.U.: Labor market concentration and heterogeneous effects on wages: evidence from Japan. J. Jpn. Int. Econ. **67**, 101242 (2023)
26. Azar, J., Marinescu, I., Steinbaum, M.: Labor market concentration. J. Human Resour. **57**(S), S167–S199 (2022)
27. Principles of Economics University of Minnesota Libraries Publishing (2016)
28. Dempere, J.M., Pauceanu, A.M.: The impact of economic-related freedoms on the national entrepreneurial activity. J. Innov. Entrepr. **11**(1), 1–20 (2022)
29. Miller, B., Atkinson, R.D.: Digital drag: ranking 125 nations on taxes and tariffs on ICT goods and services. Inf. Technol. Innov. Found. (ITIF) (2014)
30. Li, H., Wu, D., Chen, J., Chan, K.C.: The impact of technology export regulations on corporate R&D investments. Borsa Istanbul Rev. **23**(2), 322–333 (2023)

31. Zhang, H., Wei, Y., Ma, S.: Overcoming the "Solow paradox": tariff reduction and productivity growth of Chinese ICT firms. J. Asian Econ. **74**, 101316 (2021)
32. Liu, J., Guo, Z.: Effects of labor market flexibility on foreign direct investment: China evidence. Financ. Res. Lett.. Res. Lett. **53**, 103574 (2023)
33. Ali, M.S.B.: Does ICT promote democracy similarily in developed and developing countries? A linear and nonlinear panel threshold framework. Telemat. Inform. **50**, 101382 (2020)
34. Erkut, B.: Entrepreneurship and economic freedom: Do objective and subjective data reflect the same tendencies? Entrepr. Bus. Econ. Rev. **4**(3), 11–26 (2016)
35. Gilder, G.: Metcalfe's law and legacy. Forbes ASAp, **13** (1993)
36. Schoder, D.: Forecasting the success of telecommunication services in the presence of network effects. Inf. Econ. Policy **12**(2), 181–200 (2000)
37. Varian, H.R.: Economies of Information Technology. University of California, Berkeley, Mimeo (2001)
38. Aoun, D., Hwang, J.: The effects of cash flow and size on the investment decisions of ICT firms: a dynamic approach. Inf. Econ. Policy **20**(2), 120–134 (2008)
39. Burton, M.D., Sørensen, J.B., Beckman, C.M.: Coming from good stock: career histories and new venture formation. In: Social Structure and Organizations Revisited, pp. 229–262. Emerald Group Publishing Limited (2002)
40. Davila, A., Foster, G., Gupta, M.: Venture capital financing and the growth of startup firms. J. Bus. Ventur. **18**(6), 689–708 (2003)
41. Lasch, F., Le Roy, F., Yami, S.: Critical growth factors of ICT start-ups. Manag. Decis.Decis. **45**(1), 62–75 (2007)
42. Eurostat: https://ec.europa.eu/eurostat/databrowser/view/isoc_sks_itspt/default/table?lang=en. Last accessed 15 May 2023
43. BoldData: https://bolddata.nl/en/. Last accessed 15 May 2023
44. Bureau of Labor Statistics; Economic Research Institute: https://codesubmit.io/blog/software-engineer-salary-by-country/. Last accessed 15 May 2023
45. EF Education First: https://www.ef.com/wwen/epi/. Last accessed 15 May 2023
46. International Telecommunication Union (ITU) World Telecommunication: https://data.worldbank.org/indicator/IT.NET.USER.ZS. Last accessed 15 May 2023
47. UNESCO: https://rankedex.com/society-rankings/education-index. Last accessed 15 May 2023
48. NordLayer: https://nordlayer.com/global-remote-work-index/. Last accessed 15 May 2023
49. Fragile States Index: The New Humanitarian, https://fragilestatesindex.org/global-data/. Last accessed 15 May 2023
50. United Nations Department of Economic and Social Affairs: https://www.worlddata.info/average-age.php. Last accessed 15 May 2023
51. The Heritage Foundation: https://www.heritage.org/index/pdf/2021/book/2021_IndexofEconomicFreedom_Highlights.pdf. Last accessed 15 May 2023
52. Trading Economics: https://tradingeconomics.com/country-list/corporate-tax-rate?continent=asia. Last accessed 15 May 2023
53. Hofstede Insights: https://clearlycultural.com/geert-hofstede-cultural-dimensions/individualism/. Last accessed 15 May 2023
54. OECD: https://stats.oecd.org/Index.aspx?DataSetCode=VC_INVEST. Last accessed 15 May 2023
55. Sipotra.it: https://www.sipotra.it/wp-content/uploads/2019/04/ICT-INVESTMENTS-IN-OECD-COUNTRIES-AND-PARTNER-ECONOMIES.-TRENDS-POLICIES-AND-EVALUATION.pdf. Last accessed 15 May 2023

Application of Immersive Technologies to Improve the Effectiveness of Training and Staff Performance

Olga Rostova⬡, Svetlana Shirokova⬡, and Anastasiia Shmeleva⬡

Abstract The chapter discusses the possibilities of using immersive approach for training employees of oil and gas enterprises realized through virtual and augmented reality technologies. The study presents a project of introducing virtual and augmented reality technologies in the processes of training employees of repair teams and performing repairs of downhole equipment, which is planned to be implemented in the company PJSC Gazpromneft. The author describes the prerequisites of the project initiation and peculiarities of immersive technologies use for solving specific problems arising in the training of specialists in this field. Based on the author's methodology the project management methodology is substantiated, the digital innovation potential of the organization is assessed and recommendations for its improvement are given. Potential barriers that may arise in the implementation of technologies, expected effects and results are determined. It is important to identify promising directions for the use of immersive technologies in oil and gas companies.

Keywords Virtual and augmented reality technologies · Immersive learning approach · Digital projects · Innovative potential of the organization · Flexible management methods · Performance evaluation

1 Introduction

Immersive technologies are currently at the initial stages of their development, and their possibilities in various spheres are not always obvious yet. Nevertheless, they have already found serious application in the processes of employee training and are used to improve the efficiency of enterprise personnel [1].

O. Rostova · S. Shirokova (✉)
Peter the Great St. Petersburg Polytechnic University, Saint-Petersburg, Russia
e-mail: swchirokov@mail.ru

A. Shmeleva
Marriott International Yerevan, Yerevan, Armenia

Immersive technologies are a set of methods and tools for immersing the user in a virtual environment using the impact of audio-visual effects and specialized content on the user's senses. Currently, immersive technologies are primarily virtual and augmented reality technologies [2].

Virtual reality is a three-dimensional digital environment generated with the help of special programs and technical means, transmitted to the user through his senses and feelings, representing the likeness of the surrounding real world.

Augmented reality is a direct or indirect representation of the real environment, the elements of which are "supplemented" with digital information or objects using computer graphics, audio and other sensory data in real time.

The presented research considers the project of introducing virtual and augmented reality technologies in the processes of training employees of repair teams and performing repairs of downhole equipment, which is planned to be implemented in the company PJSC Gazpromneft.

In the oil and gas industry, the decline in the quality of oil reserves, the focus on hard-to-recover reserves, and the trend towards renewable energy lead to higher costs and lower profits in the production chain of oil companies. A significant problem is the downtime of oil and gas equipment due to accidents and lost profits during these downtimes [3]. In this regard, there is a constant search for new technological solutions that can improve efficiency and reduce costs. One of such solutions may be the use of virtual and augmented reality technologies to improve the efficiency of personnel work [4–7].

Public Joint Stock Company (PJSC) Gazpromneft has become one of the most prominent Russian industrial companies working with AR/VR technologies. The company has established the AR/VR Technology Competence Center, which is an internal aggregator aimed at improving organizational readiness for AR/VR implementation. The most significant areas that the company plans to develop in the coming years are presented in Table 1.

In some areas, projects are already underway, both pilot launches and full-scale project deployments. Some of the projects are still at the planning stage, including the project on training of personnel operating complex equipment, which is further discussed in detail in the study.

The aim of the study was to analyze the peculiarities of using an immersive approach for training employees of oil and gas enterprises, implemented through virtual and augmented reality technologies, as well as the possibilities of implementation and effectiveness of the presented digital project.

2 Materials and Methods

As a methodological basis of the research the works [8–10], as well as studies related to the use of digital technologies in the field of personnel training [1, 11] and increasing the human resource potential of organizations [12, 13] acted as a methodological basis for the research. The issues of planning and realization of immersive technology implementation projects, as well as the peculiarities of evaluating their effectiveness are devoted to the works [14–16].

The following methods were used to justify the project:

Table 1 Areas of AR/VR utilization in Gazpromneft Company

Technology	Usage directions	Goals	Effects	Barriers	Result
VR	Personnel training	Each subsidiary is equipped with a VR classroom	70% reduction of risk probability 50% reduction of training costs	Comfortable time in VR not more than 30 min Lack of infrastructure Need for content creation Insufficient level of immersion (quality of content realization)	Corporate VR management environment (integration of all VR content created in the company into the VR course builder)
VR	Working with engineering models	90% of engineering models are accepted in VR	A 10% reduction in design time	Lack of infrastructure in subsidiaries and design institutes Need to convert engineering models	Automated converter of engineering models to VR
AR	Construction control	80% of objects are controlled using AR	By 7% reduction of construction time 30% reduction in the cost of construction control	Limited number of devices (AR glasses) supporting scenarios Limited operating time of AR-glasses Need for model conversion	Construction control system (AR module)

(continued)

Table 1 (continued)

Technology	Usage directions	Goals	Effects	Barriers	Result
AR	Control of equipment operation	Use of interactive maintenance instructions for 80% of equipment	Reduction of equipment maintenance costs by 30%	Limited number of technologically suitable devices (certified for explosive objects), their high cost Diverse fleet of equipment, difficulty in scaling	IT solution for management AR-workflow
AR	Warehouse Operations Support	Equipped with AR glasses 80% of warehouses	Reduction of time and labor intensity of warehouse operations by 20%	Necessity of integration with other warehouse information systems	Warehouse operations management system (AR module)

1. The author's methodology was used to select a methodology for digital project management, which allows us to form a project pattern and offer recommendations based on the peculiarities of its implementation [17].
2. To assess the digital innovation potential of the organization necessary for the implementation of immersive technologies and justify the directions of its adaptation, taking into account the tasks set in the project, the author's methodology of analysis of matrix sets of "Strengths and Weaknesses", "Opportunities and Threats" was used [18].
3. The method of real options was used in assessing the effectiveness of the project. The validity of the application of this method is determined by the conditions in which the project is realized: on the one hand, it is a high degree of risk and the likelihood of obtaining new information during the project implementation, on the other hand, the availability of the management of the ability to respond flexibly, that is, to change the direction of the project and the execution of real options [19].

3   Results

Gazpromneft-Digital Solutions, which is planning the project, is a subsidiary of PJSC Gazpromneft, is engaged in the implementation of projects using digital technologies, and is represented in 30 regions of Russia.

The aim of the project was to improve the efficiency of downhole workover processes at a selected group of wells by introducing virtual and augmented reality technologies into the processes of training of the workover crews' employees and performing repairs of downhole equipment.

Three possible directions of using the development were considered:

1. Development of simulators and trainers for training operators and repairmen of complex process equipment.
2. Repair and maintenance of downhole equipment, checking its serviceability and ensuring the quality of repair work.
3. Digital modeling of equipment and structures for repair works, work with engineering models.

In the basic scenario of the project only the development of VR-simulators and simulators for employee training was envisaged, the second and third areas listed above can be implemented later depending on the success of the basic development.

As a result of implementation of AR/VR systems it is possible to obtain the effects presented in Table 2.

If the project is expanded to the use of augmented reality technologies in the repair process, it will allow employees working with equipment to access technical documentation and quickly contact support specialists right at the site of work.

In a project scenario involving digital modeling of equipment and structures for repair work, design time will be significantly reduced. Acceptance of models in virtual reality reduces the time, number of operations and iterations when interacting with the design institute, improves the quality of models, and reduces the cost of producing laboratory samples of equipment and structures.

The author's algorithm presented in [17] was used to select the most appropriate project management methodology.

At the first stage, the degree of project uncertainty was assessed using the Stacey matrix. The axes of the matrix reflect the degree of project uncertainty in two directions-customer requirements and the method of project realization. In both directions the project is characterized by a high level of uncertainty, because the customer formulated the main objectives of the project, but more detailed requirements will be formed and changed in the course of the project.

The way of realization of the set tasks at the initial stage is also not fully defined, because the project on development of VR-system for training of operators and repairmen of complex technological equipment is quite new not only for the company itself, but also for the industry as a whole. Currently, mainly pilot projects are being implemented in this direction. In this regard, in accordance with the Stacy matrix, it is recommended to use flexible and hybrid methodologies to manage such projects.

At the next stage, a project pattern was built to assess its complexity. The results of the analysis confirmed the feasibility of using a hybrid methodology that combines the tools of agile and classical project management (Fig. 1). In particular, waterfall methodology is used to build a hierarchical structure of project work (HSPW) that allows to decompose the work into smaller components, while agile methodology is used for product development and testing. Other branches of the ISR such as pre-project survey, equipment supply, user training are implemented according to the waterfall model.

Table 2 Effects from implementation of AR/VR systems

Project	Effects
Development of simulators for training operators and repairers of complex technological equipment	1. Reducing the shortage of qualified personnel by shortening the period of practical training of employees to work with downhole equipment 2. Improving the quality of personnel training, regular testing of skills and strengthening of knowledge in preparation for certification; 3. Transparent assessment of the brigade's readiness to carry out operational activities due to detailed statistics on the level of training of each participant 4. Development of personnel motor skills and thinking strategies in non-routine situations in a safe environment 5. Optimization of training effort by practicing skills independently without the need for a mentor 6. Reduce the cost of purchasing physical samples of equipment for personnel training
Repair and maintenance of downhole equipment, checking its serviceability and ensuring the quality of repair work	1. Reduce equipment downtime through faster and more effective incident response 2. Reduction of equipment maintenance errors due to the possibility of consulting a specialist carrying out work with experts 3. Increased safety of repair operations 4. Increased productivity of repair specialists due to the possibility of obtaining the necessary technical information in real time 5. Cost reduction due to the absence of the need for personal presence of a wide range of specialists to carry out repairs
Digital modeling of equipment and structures for repair work, work with engineering models	1. Acceptance of models in virtual reality reduces the time, number of operations and iterations when interacting with the design institute 2. Significant reduction in overall design time 3. Improved quality of models of equipment and structures for repair work 4. Reduction of costs for manufacturing laboratory samples of equipment and structures

Hybrid methodology allows to ensure the basic principles of project management: delivery of project results on time with iterative and dynamic planning, ensuring quick response to flexible requirements and effective decentralization to ensure the required product quality. The hybrid methodology covers all levels of digital project management: senior management controls the project by key milestones, the project

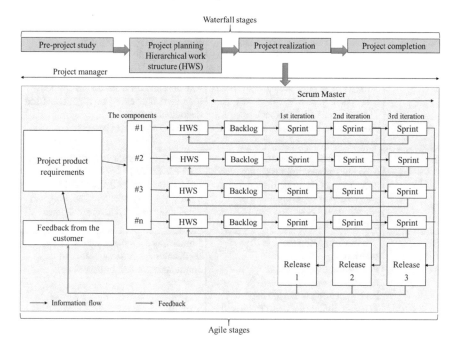

Fig. 1 Project management scheme using hybrid methodology

manager is responsible for the deliverable results, which are provided by the project team using an iterative approach, providing quality feedback to customers.

Further, based on the author's methodology for assessing the digital innovation potential of an organization, as well as the analysis of sources [20–24], the factors that are most important for the implementation of the project for the introduction of virtual and augmented reality technologies were identified. The selection of factors was made taking into account the perspectives of the balanced scorecard system, i.e. the lists of factors include elements characterizing the level of development of digital innovation potential according to different perspectives.

Experts were asked to assess the factors of each group on a five-point scale. The results of the survey made it possible to assess the level and significance of the strengths and weaknesses of the organization's digital innovation potential for effective project implementation, as well as the influence of the factors of opportunities and threats on the formation of digital innovation potential. In order for the experts to unambiguously understand the correspondence of the level of factors to the ballot assessment, a description of possible variants of the levels was compiled. A fragment of the description for assessing the level of force factors is presented in Table 3. After each stage of the assessment, tables of results were compiled, and the degree of consistency of experts' opinions on all four groups of factors was determined.

At the next stage, the cost-effectiveness of the project of implementing virtual and augmented reality technologies using the real options method was assessed

Table 3 Fragment of a table for assessing the level of factors in the realization of a digital project

Assessment	Readiness of the organization to change business processes to meet the challenges of digitalization	Having a well-established process for creating new digital products
1.	There is no understanding of the basic concepts of process management There is no unified approach to process description and process model There are no process owners	Products and services are not explicitly identified, they exist in a fragmented and disjointed manner There are no product owners
2.	Description of processes in the form of different parts in different normative documents of the organization, there are no rules for making changes An automated process control system has been selected	A portfolio of products and ser-services has been created Product owners are defined and authorized. Key rules for designing and developing new products are defined
3.	The modeling notation has been selected Process catalog is created Process owners are defined Re-engineering is carried out The first digital products are created	Implemented a system for collecting feedback from product users and continuous product improvement
4.	At least 30% of business processes are automated Fragmentary monitoring of indicators and efficiency of business processes is performed	New product development processes are planned digitally from the outset Product owners realize the full potential of development teams
5.	The process management system is implemented throughout the organization as an integrated process monitoring and optimization modeling tool	The process of creating digital products is fully established, easy-to-use development tools, end-to-end integration of development and maintenance processes

[25]. Extensive calculations on the components of investment and operating costs in the implementation of VR/AR projects, as well as the economic effect of their implementation, are made on the basis of the analysis of studies conducted by the international consulting IT-company CapGemini, and the Russian analytical center TAdviser together with the company "CROC Immersive Technologies", specializing in the application of immersive technologies in industrial industries [26].

In the structure of investment costs in projects based on VR/AR technologies, in addition to the components standard for IT projects, there are specific areas of costs, such as payment for the development of content and scenarios for the virtual or augmented reality environment. According to the research results, the ratio between the components of investment costs depends on the type of VR/AR project (see Table 4).

Any change in simulated reality, to avoid false skill, should be reflected in the VR/AR content. The main economic effects resulting from the implementation of

Table 4 Ratio of investment cost directions depending on the type of VR/AR project

Areas of expenditure	Training staff to operate complex equipment (%)	Distance education system (%)	Equipment repair and maintenance (%)	Virtual models of industrial and infrastructure facilities (%)	A comprehensive VR simulator for dealing with emergencies (%)
Specialized hardware	50	2	40	10	10
Specialized software	10	10	10	15	5
Payment for integration work	10	28	10	5	5
Payment for content development	20	45	30	55	65
Payment for work on script development	10	15	10	15	15

virtual and augmented reality technologies for each of the considered projects are presented in Table 5.

The chapter considers the possibility of applying three types of real options for the project.

1. Development option-for the project under consideration it is an opportunity for further development by including an additional line of activity into the project. In addition to personnel training using virtual reality technology, the possibility of developing an augmented reality program complex for oil and gas equipment maintenance and repair work is considered.
2. Option to expand the scale of the project is an option-call, which gives the opportunity, if the developed technology is successfully implemented for one object (pilot project), to invest additional funds and scale the project to two other objects.
3. Switch option is an option that gives an opportunity, in case of unfavorable development of events, to change the final product of the project and reorient it to digital modeling of equipment and structures for repair works.

Table 6 summarizes all the NPV results obtained considering the use of options and calculates their value to the project.

The use of real options will increase the value of the project under consideration by 6,291,309 rubles, increasing the value of net present value of the project to 8,237,166 rubles. And the value of return on investment to increase from 1.04 to 1.19, which allows to attribute this project to the category of effective (efficiency criterion PI > 1.15).

Table 5 Economic effects resulting from the implementation of virtual and augmented reality technologies

Project	Change of indicators	Economic effect
1. Development of simulators for training operators and repairers of complex technological equipment	Additional revenue	Revenue losses are reduced by reducing accidents that occur due to the low level of training of employees responsible for complex equipment
	Cost reduction	Reducing the cost of purchasing expensive equipment that is used in the training process, and the salaries of employees providing training, since with independent training the need for them will be reduced
2. Repair and maintenance of downhole equipment, checking its serviceability and ensuring the quality of repair work	Additional revenue	Reduction of revenue losses due to long equipment downtime during repair and maintenance due to faster and more efficient response to incidents and increased productivity of repair specialists Reduced damage from reduced product quality
	Cost reduction	Reduced cost of transporting a wide range of specialists to remote sites Reduced costs of dealing with the consequences of poor quality repairs and equipment maintenance errors
3. Digital modeling for repair works, work with engineering models	Additional revenue	Reduced revenue loss through the use of innovative and modernized equipment and designs for repair work
	Cost reduction	Reduction of costs for manufacturing laboratory samples of equipment and structures

Table 6 Evaluating the impact of options on project performance

Option Exercise	Return on investment	NPV, rub	Option value, rub
Without options	1,04	1 945 857	
With the use of a development option	1,16	7 324 121	5 378 264
With the use of a switch option	1,06	2 858 902	913 045
With the use of a complex option	1,19	8 237 166	6 291 309

4 Discussion

It should be noted that companies realizing projects based on virtual and augmented reality technologies have to face a number of barriers to achieve meaningful results. A study of a number of sources [27–29] made it possible to identify the most significant of them:

1. Complexity in the interaction between developers and industry customers. To implement industry-specific solutions, developers need to know not only the technological capabilities of AR/VR, but also the specifics of the tasks of a particular industry for which the solution is being developed. On the part of customers there is a difficulty in understanding the potential opportunities that can be provided by the use of virtual and augmented reality technologies and, as a consequence, difficulties in the correct formation of development tasks.

2. The application of immersive technologies should be based on a prepared technological foundation. Several factors play a role here. Insufficient level of development of sub-technologies for realization of a specific technological solution is possible. Lack of access to specialized foreign equipment and software due to sanctions and import substitution policy, provided that domestic analogues of the required quality are not available on the market. As well as technological unpreparedness of the enterprise itself.

3. Lack of qualified specialists. Almost all large companies have their own competence centers and teams that prepare solutions based on immersive technologies. However, even in such centers it is not always possible to gather full-cycle specialists who would be able to solve problems of any level of complexity.

5 Conclusions

The study analyzed promising areas for the use of VR/AR virtual and augmented reality technologies in PJSC Gazpromneft, identifying potential barriers that may arise in the implementation of technologies, expected effects and results.

On the basis of using the algorithm of project management methodology selection it was determined that the project management of immersive technologies implementation in the processes of repair crews staff training and well intervention equipment repair is optimally carried out using hybrid methodology. In this regard, it is recommended that in the course of the project implementation to monitor the indicators inherent in both classical project management, used in the company in accordance with the regulations, and indicators reflecting a flexible approach and allowing to assess the quality of the digital product, speed of development and compliance with customer requirements.

The application of the mechanism of adaptation of the organization's innovation potential to the tasks of its innovative development allowed us to identify the significant factors of digital innovation potential for the implementation of the project of introducing virtual and augmented reality technologies that require adaptation measures. The most significant ones include: increasing technological and organizational readiness of the enterprise, formation of cross-functional teams, increasing competencies, interest and involvement of potential users.

Embedding real options in the investment analysis of a digital innovation project will increase its net present value by 4 times and increase the return on investment by 15% in relation to the base scenario of the project without options. The envisaged possibility of using real options for development, for scaling and for switching to another final product will increase the management flexibility of the project in conditions of high uncertainty of the external and internal environment of the organization.

References

1. Dengel, A., Mazdefrau, J.: Immersive learning explored: subjective and objective factors inuencing learning outcomes in immersive educational virtual environments. In: 2018 IEEE International Conference on Teaching, Assessment, and Learning for Engineering (TALE), Wollongong, Australia, pp. 608–615 (2018)
2. Al Qrain, A., Reddy, G., Jadhav, S.: Virtual reality simulators in the oil & gas industry review of existing solutions and method for evaluation. In: Abu Dhabi International Petroleum Exhibition & Conference. Abu Dabi, UAE (2020)
3. Kirillov, D.S.: Augmented reality for industry: efficiency and methods of use. Sphere Oil Gas 2(76), 36–38 (2020)
4. Azieva, R.K., Tayamaskhanov, H.E., Zelimhkanova, N.Z.: Assessing the readiness of oil and gas companies for digital transformation. In: European Proceedings of Social and Behavioural Sciences, vol. 117, pp.1852–1862 (2021)
5. Flaksman, A., Kokurin, D., Khodzhaev, D., Ekaterinovskaya, M., Orusova, O., Vlasov, A.: Assessment of prospects and directions of digital transformation of oil and gas companies. In: IOP Conference Series Materials Science and Engineering, vol. 976, no. 1, p. 012036 (2020)
6. Linnik, Y., Kiryukhin, M.: Digital technologies in the oil and gas industry. Vestnik Universiteta 1(7), 37–40 (2019)
7. Sola, E., D'Angelo, V., Capo, F.: Digital transformation in the energy industry. In: Handbook of Research on IoT, Digital Transformation, and the Future of Global Marketing, pp. 97–111 (2021)
8. Aldea, A.: Modeling and analyzing digital business ecosystems: an approach and evaluation. In: 2018 IEEE 20th Conference on Business Informatics (CBI), pp. 156–163. Vienna, Austria (2018)
9. Baker, C., Fairclough, S.: Adaptive virtual reality. In: Current Research in Neuroadaptive Technology, pp.159–176 (2022)
10. Kozlov, A., Teslya, A.: Digital potential of industrial enterprises: essence, determination and calculation methods. Transbaikal State Univ. J. 25(6), 101–110 (2019)
11. Garcia, C., Naranjo, J., Gallardo-Cardenas, F., Garcia, M.: Virtual environment for training oil & gas industry workers. In: Augmented Reality, Virtual Reality, and Computer Graphics, LNIP, vol. 11614 (2019)
12. Lee, E.A.L., Wong, K.W.: A review of using virtual reality for learning. In: Transactions on Edutainment I. LNCS, vol. 5080, pp. 231–241. Springer, Heidelberg (2008).
13. Naranjo, J., Gallardo-Cardenas, F., Garcia, M.: An Approach of virtual reality environment for technicians training in upstream sector. IFAC-Pap. 52(9), 285–291 (2019)
14. Shirokova, S., Kislova, E., Rostova, O., Shmeleva, A.: Company efficiency improvement using agile methodologies for managing IT projects. In: International Scientific Conference on Digital Transformation on Manufacturing, Infrastructure and Service, vol. 25, pp. 1–10. Saint Petersburg, Russia (2020)
15. Zhu, F.: Digital project evaluation and development. Digitalization and Analytics for Smart Plant Performance, pp. 459–475 (2021)
16. Grozdova, A., Shirokova, S., Rostova, O., Shirokova, A., Shmeleva, A.: Rationale for information and technological support for the enterprise investment management. Lect. Notes Netw. Syst. 387, 181–190 (2022)

17. Shmeleva, A.S.: Algorithm for choosing a methodology for managing digital innovation projects. J. Manag. Res. **2**, 42–52 (2022)
18. Shmeleva, A., Suloeva, S.: Development of a mechanism for adapting digital innovationpotential of an organisation with allowance for peculiarities of digital innovation projects. Sustain. Develop. Eng. Econ. **2**, 5 (2022)
19. Rostova, O.V., Rostova, A.S., Rodionova, E.S.: Application of a method of real options in innovative projects management. Adm. Consult. **11**, 61–71 (2017)
20. Eremin, N., Sardanashvili, O.: The innovative potential of digital technologies. In: Actual Problems of Oil and Gas: Scientific Network Edition, vol. 3, No. 18, pp. 1–9 (2017)
21. Kotarba, M.: Measuring digitalization—key metrics. Found. Manage. **9**(1), 123–138 (2017)
22. Lokuge, S., Sedera, D., Grover, V.: Organizational readiness for digital innovation: development and empirical calibration of a construct. Inf. Manage. **56**(3), 445–461 (2018)
23. Panjaitan, R., Moonti, A., Adam, E.: Technology readiness and digital competing capabilities: digital value resonance. J. Econ. Bus. **6**(2), 205–226 (2021)
24. Schumacher, A., Nemeth, T., Sihn, W.: Roadmapping towards industrial digitalization based on an Industry 4.0 maturity model for manufacturing enterprises. Procedia CIRP **79**, 409–414 (2018)
25. Rostova, O., Shirokova, S., Sokolitsyna N., Shmeleva, A.: Management of investment process in alternative energy projects. In: International Science Conference SPbWOSCE-2018 Business Technologies for Sustainable Urban Development, vol. 110, p. 02032 (2019)
26. The market of industrial VR/AR solutions in Russia. TAdviser research, https://www.tadviser.ru/index.php/Статья:Рынок_промышленных_VR/AR-решений_в_России_(исследование_TAdviser). Last accessed 20 Sep 2023
27. Chanias, S., Myers, M.D., Hess, T.: Digital transformation strategy making in pre-digital organizations: the case of a financial services provider. J. Strat. Inf. Syst. **28**(1), 17–33 (2019)
28. Stoianova, O.V., Lezina, T.A., Ivanova, V.V.: The framework for assessing company's digital transformation readiness. Vestnik SPbGU **36**(2), 243–265 (2020)
29. Salko, M.: Developing the innovative potential of digital transformation of enterprises of the fuel and energy complex. Tyumen State Univ. Herald Soc. Econ. Law Research **7**(2), 200–218 (2021)

Conceptual Model of Strategic Management of the Quality Management System as a Tool for Transforming Labor Relations in the Context of Digitalization

Boris Lyamin⬤, Maxim Ivanov⬤, and Lyaukina Gulnara

Abstract The author identified the basic principles of building a quality management system at modern enterprises in the conditions of economic instability, carried out an in-depth analysis of the existing QMS management models at the enterprise, developed by domestic and foreign scientists, such as Kachalov R.M., Sleptsova Yu.A., Rogova M.V., Gusev I.S., Borgardt E.A., Yashin N.S., Andreeva T.A., Demidovets V.P. and etc. In modern times the Russian economy is undergoing significant restructuring in all areas. The supply chain of raw materials is changing, materials, finished products, the structure of energy supplies is changing, the share of resources allocated to R&D is increaing, since there are currently no analogues of some outdated technologies, it is necessary to significantly optimize the company's internal business processes. In this regard, the author has developed a conceptual model of strategic management of the quality management system, which allows taking into account environmental factors, including force majeure factors, and using the PDCA cycle to develop measures to improve the QMS to achieve the strategic goals of sustainable development of the enterprise. A feature of the developed model is its focus on the organization's personnel, whose qualifications in the current conditions are given special attention. It is the personnel who will be able to create new systems, technologies, products and services.

Keywords Quality management system · Management model · Personnel management · International resource limitations · Digital transformation

B. Lyamin (✉) · M. Ivanov
Peter the Great St.Petersburg Polytechnic University, St.Petersburg, Russia
e-mail: Lyamin_bm@spbstu.ru

L. Gulnara
Kazan State Power Engineering University, Kazan, Russia

1 Introduction

In modern conditions, the level of competition in various sectors of the economy is rapidly increasing. This trend is associated with a number of reasons, including globalization and the development of technology, which lead to the formation of unified sales markets. An increasing number of players are appearing on the market, the consumer has an increasing opportunity to choose similar goods and services from different manufacturers and suppliers, therefore, competition between companies is increasing, since it is in the interests of any manufacturer to ensure that people buy its products and not competitors. It should be noted that the priorities will also be to expand its market share, the desire not only for leadership, but also for a monopoly.

The development and implementation of digital technologies in various areas continues actively. In addition, recently more and more enterprises have been using information systems and technologies in their work, designed to optimize many processes in the organization. However, the needs and demands of consumers also increase with the development of progress. It is becoming increasingly difficult to satisfy the consumer, to offer him products that not only meet the needs of the client, but also the directions of industry development [1].

Modernization of existing and development of new digital technologies helps to optimize time and significantly simplify certain elements in operational activities. Thanks to computer technology, the ability to quickly analyze large volumes of information has become available, as well as reducing the number of errors that may occur among workers, since the human factor will play a significant role. In addition, the use of digital technologies can significantly improve communication between different departments and employees, which in turn contributes to more efficient functioning of the company as a whole. At the same time, information technology can also cause increased competition, therefore, organizations have to develop their strategies more carefully [2].

Strategic management is a set of activities that allows you to realize the long-term goals of an organization in a changing external environment. During this management, significant decisions for the company are formed and made that can change the existing mission and strategy of the enterprise.

The main direction of strategic management will be the long-term perspective, that is, the organization's goal is not just to sell products and make a short-term profit, but to move and develop further, expand sales markets, customer base, etc. At the same time, only existing methods of business development are not used; there is also an active search for new options for improvement and modernization. With this type of enterprise management, an analysis will be made not only of the internal environment, but also the external one, due to which the developed solutions and strategies will be based on more complete and objective data, since in addition to the analysis of the organizational structure and processes occurring within the company, information about opportunities will be taken into account and threats existing in the market, contact audiences, suppliers, etc. [3, 4].

It should be emphasized that strategic management and management based on long-term planning are not identical. The main difference is that long-term planning is formed from what is happening in the present to what is planned in the future, and strategic planning comes from what the organization wants to have in the future and how it can get to this "future", based on what it has at the moment, taking into account changing conditions. When drawing up various development scenarios (forecasting), strategic planning will involve the assessment and subsequent preparation of forecasts not only of the external, but also the internal environment of the organization, while in long-term planning, extrapolation of current trends is carried out. Thus, in strategic management there is a constant determination of what the organization must do now in order to achieve its desired goals in the future, taking into account that the environment and circumstances in which the organization operates can and will change, i.e. Strategic management is a kind of view of the present from the future.

Employees in strategic management act as a key resource of the enterprise, a guarantee of its sustainable development, in contrast to operational management, in which the staff is only an executive unit that produces the company's goods and services. If the organization has a development strategy, management is interested in investing in its employees, understands why to train people, invest their resources in them, since qualifications, competencies, skills, experience, as well as motivation have a serious impact on the functioning of the enterprise, achieving goals and implementation strategies. It should be noted that when changing strategies, as well as introducing new technologies, quite often you have to face resistance from your employees. For this reason, another important component of strategic management will be the competent construction of a personnel management system, primarily motivation. This is especially true for service sector enterprises, where not only the quality of the service provided, but also the client's impressions and the desire to use the services again will depend on the staff. This affects work in the hotel and tourism sectors to a greater extent; this is due to the high communication load of staff and constant interaction with guests. In this case, insufficient motivation of employees leads to a decrease in the quality of service and an increase in staff turnover. In addition, dissatisfaction and disinterest in employees are visible to customers, which also negatively affects the experience of service consumers.

Strategic management is carried out by the top-level management of the organization (administrative and managerial personnel), decisions are made depending on certain tasks facing the organization and its leaders. In this case, a management decision will be the result of an analysis of a problem situation, the determination of ways to resolve it, and then the choice of a specific path that will be used in a particular case, and the tools with which the solution will be implemented. It is important to take into account that strategic management is characterized by the presence of variability in decisions, since the cost of an error can be very high and even destructive for the enterprise. For this reason, several alternative options for both the development of events and ways to respond to them should be proposed [5].

Thus, in the context of the difficult economic and political situation in the world, enterprises need not only to constantly monitor and analyze the current situation in the internal and external environment, but also to assess prospects in conditions of uncertainty, calculate various options, assess possible risks and probabilities their occurrence, as well as the degree of their negative impact on the company, strive to improve existing technologies, processes and introduce more and more new and effective ones.

2   Materials and Methods

During the implementation of the research work, methods of analysis, synthesis, modeling, generalization of scientific works and identification of the most promising methodological aspects that have the greatest potential for practical application were used.

To conduct the research, the works of domestic and foreign scientists involved in the creation of methodological approaches to the strategic management of enterprises were analyzed. Reporting documents on the construction of quality management systems of large domestic enterprises in the oil and gas industry, the service sector and information technology were used. Based on the data obtained, using theoretical research methods, the basic principles of building a quality management system in the context of digital transformation were identified. Since existing quality management system management models do not take into account the current macroeconomic situation and changes in work with suppliers and partners, the modeling method was used and a conceptual model of strategic management of the quality management system at enterprises in the context of digital transformation was developed.

The purpose of the work is to develop a conceptual model of strategic management of an enterprise's quality management system as a tool for transforming labor relations in the context of digitalization.

To achieve this goal it is necessary:

- consider the principles of building a quality management system at an enterprise in the context of digital transformation;
- analyze existing approaches to strategic enterprise management in the context of digital transformation;
- to develop a conceptual model of strategic management of the enterprise's quality management system as a tool for transforming labor relations in the context of digitalization, allowing the use of the existing potential in the enterprise for its sustainable long-term development in conditions of instability of external economic factors.

3 Results

3.1 Principles of Building a Quality Management System in the Context of Digital Transformation

Having analyzed successfully functioning quality management systems at enterprises in various sectors of the economy, for example, mining and manufacturing enterprises, educational organizations, service organizations, etc. we can highlight the basic principles on which the most stable and effective systems are built (Fig. 1) [6–12]:

- quality is an element of safety; within the framework of this principle, a set of measures is carried out to ensure product safety, safety of workers, i.e. the quality of processes must be such as to prevent the possibility of a threat to the safety of life and health of both personnel and consumers;
- leadership—the principle of leadership implies not only providing management with conditions for effective work of employees, but also supporting local initiatives, i.e. team leaders who set an example for their colleagues through their work;
- employee involvement—allows the most complete use of the enterprise's human capital; if the enterprise's employees feel needed and useful to the company, then their labor productivity and the number of rationalization proposals can increase significantly;

Fig. 1 Principles for constructing a Quality Management System in the context of digital transformation

- focus on the consumer—for sustainable development, the enterprise constantly ensures that consumer satisfaction is at a high level, accordingly, the introduction and improvement of the quality management system allows us to most fully satisfy the consumer; we will provide him with the goods at the required time, in the required quantity at the agreed price and of the same quality what the consumer wants;
- process approach, any actions in the organization must be considered as inter-connected processes, accordingly, to increase the efficiency of activities, these processes are identified and optimized;
- a systematic approach allows you to manage the organization at the management level, taking into account the long-term goals of the enterprise and the designated principles of quality management;
- constant improvement, oil and gas sector enterprises operate in conditions of fierce competition in the global energy market, and accordingly, in order to with-stand competition, they need to constantly improve their activities, i.e. constantly improve business processes;
- making decisions based on evidence, this principle is especially important in conditions of uncertainty, when the external environment has high volatility; making informed management decisions based on confirmed information can significantly increase the competitiveness of the company;
- relationship management, this principle is of particular importance for oil and gas companies, since the main consumers are foreign companies; in such economic conditions, the satisfaction of all stakeholders is not a trivial task, the solution of which will allow the enterprise to develop sustainably in the conditions of digital transformation and international resource constraints.

3.2 Analysis of Existing Approaches to Strategic Enterprise Management in the Context of Digital Transformation

The macroeconomic situation is currently difficult to predict, domestic enterprises are forced to rebuild business processes and optimize production based on the instability of foreign economic factors, some of the most affected by this are oil and gas industry enterprises, which, on the one hand, depend on world energy prices and are forced to reorient themselves to new ones markets, as a result of a decrease in supplies to the European Union, on the other hand, it is necessary to increase the efficiency of energy production and processing through the introduction of domestic digital technologies, since access to Western technologies is difficult. In this regard, there is a need for methods of strategic management of the quality management system. Domestic and foreign scientists consider the problem differently; various scientific schools offer methodologies for strategic management of the quality management system based on existing approaches and various levels of management. Let's look at existing approaches in more detail.

Thus, scientists of the Central Economics and Mathematics Institute of the Russian Academy of Sciences Kachalov R. M. and Sleptsova Yu. A. proposed a conceptual model of strategic management of an industrial enterprise in which the external environment plays a pivotal role [13].

The authors conduct a deep theoretical analysis of the problems of sustainable management of an enterprise and come to the conclusion that for sustainable development, an enterprise needs to constantly search for opportunities, as well as level out factors that adversely affect the activities of the enterprise. Thus, the authors pay the greatest attention to the analysis of the external environment of the enterprise and at each stage of management (intentional, expectation, cognitive, functional) it is necessary to carry out activities that will help the enterprise realize the possibilities of the external environment, while relying on the enterprise's own resources to ensure the sustainability of the enterprise in unstable conditions external environment.

The author's concept of a mechanism for managing the sustainable development of an enterprise was proposed by a scientist from the Belarusian State Technological University Rogova M. V. [14] The conceptual diagram of the mechanism for managing the sustainable development of an industrial enterprise is based on the Deming cycle (Plan-Do-Act-Act), which involves the constant improvement of business processes of an industrial enterprise for sustainable development.

Achieving the goals of sustainable development of an industrial enterprise is assumed by meeting the expectations of stakeholders, as well as implementing the criteria for sustainable development: economic (achieving economic efficiency of the enterprise in the long term), environmental (systemic measures to reduce the impact on the environment, ensuring environmental safety of production, environmentally friendly production safe products), social (creating comfortable working conditions, involving employees in enterprise management processes, using the creative potential of employees to intensify work on rational proposals). Thus, Marina Valerievna, in the proposed conceptual mechanism for managing the sustainable development of an industrial enterprise, proposes to concentrate on the internal processes of the organization, taking into account the needs of stakeholders.

An integrated approach to the problem of sustainable development of an industrial enterprise was proposed by I. S. Gusev in the work "Scientific and methodological support for multi-purpose management of sustainable development of an industrial enterprise [15] and E. A. Borgardt in his work "Strategic management of sustainable development of an enterprise" [16]. The peculiarity of the developed by I. S. The Gusev mechanism is its applicability in an unstable macroeconomic situation. According to the mechanism, achieving high levels of production and sales of products will depend on the effective organization of the performance management system; accordingly, both factors of the organization's internal environment and external factors are simultaneously taken into account. At the same time, the sustainable development of the enterprise will be expressed in the effective operation of the following types of enterprise activities: investment management, financial management, resource management, innovation management, potential management, quality management, organizational structure management, risk activity management.

Thus, a comprehensive mechanism for managing the sustainable development of an industrial enterprise involves achieving the sustainable development of an industrial enterprise through a comprehensive assessment of the enterprise's internal resources and environmental factors and operational activities developed on their basis for the main eight types of sustainable development of an industrial enterprise.

Professor Yashin N. S. and associate professor Andreeva T. A. propose to use the Hoshin Kanri methodology in the strategic management of an oil and gas corporation [17].

The Japanese Hoshin Country methodology allows the PDCA cycle to be applied at various levels of the organization's hierarchy. This methodology is especially relevant for large vertically integrated companies. Thus, according to the Hoshin Country methodology, a long-term development strategy is formed at the state level, presented in the form of a PDCA cycle, where at the "Plan" stage, based on research, a strategy for the long-term development of the oil and gas industry is formed, at the "Do" stage the strategy is implemented, at the "Check" stage "The implementation of strategies is analyzed and the strengths and weaknesses of the strategy are identified; accordingly, at the "Act" stage, corrective measures are implemented to implement the strategy for the long-term development of the oil and gas industry. Based on the developed long-term strategy, a strategy is formed at the level of the Ministry of Industry and Trade, then a medium-term strategy at the corporate level, for example, Gazprom or Rosneft, then an annual strategy at the level of a specific enterprise within the framework of corporation, finally based on the strategy of a specific enterprise the tactics of a specific unit are formed. The resulting integrated strategy for sustainable development forms a constantly improving hierarchical system that allows you to effectively manage the sustainable development of an enterprise in the context of the sustainable development of the state and industry, taking into account the requirements of consumers and stakeholders.

A scientist from the Penza State University of Architecture and Construction, as part of his work "Process management of quality management systems as part of the formation of a competitive strategy for enterprises in the construction industry" [18], developed a model of quality management in the long term as part of the development strategy of an enterprise in the construction industry.

The authors form four vectors of action to ensure effective management of product quality levels. At the first stage, it is necessary to determine the needs of the target audience and determine the company's capabilities to meet them, then determine the current level of product quality using mathematical tools; based on two measures taken, the need for material and technical renovation of the enterprise is formed; finally, the fourth vector is the organization's personnel, who directly participates in the value chain and shapes the level of product quality.

The mechanism for developing an enterprise development strategy, proposed by V. P. Demidovets, seems interesting [19]. The author proposes to increase the efficiency of the company through changes in the structure of business management and the introduction of network organizations. A network structure can be created through the development of information systems that allow, on a contractual basis, to integrate the work of suppliers of raw materials and forest processing enterprises in order to save costs and promote products to foreign markets through the use of well-known information platforms.

Thus, the considered approaches to managing the strategic development of quality systems are characterized by the level of management, tools and elaboration. In the context of digital transformation and international resource constraints, the process of managing a quality management system in the context of sustainable development of an enterprise is very relevant; for this it is necessary to formulate an approach to the strategic management of a quality management system using lean manufacturing tools.

3.3 Conceptual Model of Strategic Management of the Quality Management System at an Enterprise in the Context of Digital Transformation

Having analyzed existing approaches to strategic management of an enterprise in the context of digital transformation, we will formulate a conceptual model of strategic management of the quality management system in an enterprise (Fig. 2).

The developed conceptual model is aimed at achieving the long-term goals of the enterprise in the field of quality, taking into account the environmental factors affecting the quality management system, such as the requirements of stakeholders, macroeconomic instability, international resource constraints, as well as the need to introduce information technologies due to the digital transformation of the domestic economy. At the same time, the Quality Management System must be considered as a self-improving system operating according to the Deming cycle (PDCA).

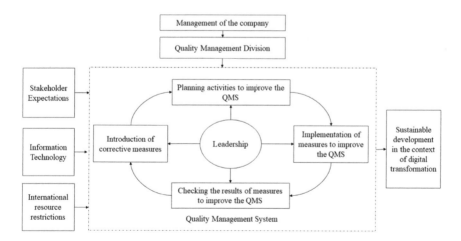

Fig. 2 Conceptual model of structural management of an enterprise's quality management system in the context of digital transformation

Let's take a closer look at the formed conceptual model.

A quality management system is necessary to improve the efficiency of an enterprise, optimize business processes, and increase the satisfaction of consumers and stakeholders. At the same time, the quality management system operates on the basis of the principle of "continuous improvement"; accordingly, the goals and objectives that the QMS faces must be constantly reviewed and improved. For this, a toolkit based on the Deming cycle is proposed.

Proposed measures to improve the QMS, improve the quality of goods or services, optimize business processes, reduce costs or increase revenue must be planned, have a clear and understandable implementation schedule and specific deadlines, as well as have the necessary resources for implementation and responsible persons. This stage is called Planning.

The next stage is called Implementation. As part of this stage, measures to improve the QMS are directly implemented at the enterprise; at this stage, it is important that the implementation schedule, resources and support from management are fully observed.

At the third stage, a comparison of the planned effects of implementation with actual results is made. It is important not only to identify deviations in the positive and negative directions during the implementation of the event, but also to identify the cause of this deviation. This stage is key in this cycle, since within this stage the factors that have the greatest impact on the implementation of measures to improve the QMS are identified.

After identifying deviations from the planned results, corrective actions are formulated that make it possible to achieve better results from the activities being implemented. During the implementation of corrective measures, new proposals for improving the QMS appear and the cycle closes. Let us formulate it as follows: "Planning—Implementation—Checking—Implementation of corrective measures." However, this cycle will not work effectively without the principle of "Leadership". Basing the PDCA cycle on the principle of "Leadership" makes it possible to actively involve the company's management in improving the QMS, who are interested in increasing the efficiency of the company's activities, since this directly affects their well-being. At the same time, the principle of "Leadership" motivates local workers who are given the opportunity to be formal or informal leaders and personally demonstrate what actions need to be taken in order to be noticed and provided with opportunities for personal and career development.

Currently, many domestic enterprises are faced with international resource restrictions from Western countries, which include, among other things, restrictions on the import of high-tech goods necessary for the economic activities of companies in various fields of activity. At the same time, this situation allows large corporations that have their own engineering centers to intensify work in cooperation with leading scientific centers in Russia and other countries to create the necessary equipment and materials.

Finally, information technology, the situation is close to the situation with Western equipment and materials. It is necessary to introduce information technologies, but suitable software is currently unavailable; therefore, it is necessary to switch to domestic programs that create, test and improve domestic IT companies.

4 Discussion and Conclusion

The research reflects significant changes in the macroeconomic situation and focuses on the need to restructure companies' business processes in conditions of uncertainty and complexity of strategic planning. At the same time, it is possible to identify the main trends that will allow enterprises to develop sustainably in the existing economic conditions:

- the most complete and effective use of human capital;
- active implementation of modern technologies, including information and digital technologies;
- reducing losses in the company's processes using modern quality management and lean manufacturing tools;
- building mutually beneficial relationships with suppliers, reducing the impact of international resource restrictions on the company's activities;
- systematic monitoring of changes in customer needs in the context of a decrease in effective demand, inflation expectations and sanction restrictions, etc.

The identified trends need to be studied in more detail, models and interaction mechanisms must be built for the most effective management of enterprises in a difficult to predict external environment.

Thus, an analysis of effectively working quality management systems was carried out, on the basis of which the basic principles of constructing a QMS were described. The approaches of domestic and foreign scientists on the strategic management of QMS were also examined in detail. Taking into account the complex macroeconomic situation and profound changes in internal processes at enterprises, a conceptual model for managing the quality management system was developed, which allows taking into account environmental factors, including force majeure factors, and applying the PDCA cycle to develop measures to improve the QMS to achieve strategic goals enterprises.

References

1. Barykin, S., Kapustina, I., Valebnikova, O., Valebnikova, N., Kalinina, O., Sergeev, S., Camastral, M., Putikhin, Y., Volkova, L.: Digital technologies for personnel management: implications for open innovations. Acad. Strateg. Manag. J. **20**, 1–14 (2021)
2. Barykin, S., Kapustina, I., Kirillova, T., Yadykin, V., Konnikov, Y.: Economics of digital ecosystems. J. Open Innov. Technol. Mark. Complex. **6**(4), 1–16 (2020)

3. Herzallah, A., Gutierrez-Gutierrez, L., Munoz-Rosas, J.F.: Quality ambidexterity, competitive strategies, and financial performance: an empirical study in industrial firms. Int. J. Oper. Prod. Manag. **37**(10), 1496–1519 (2017)
4. Konadu, R., Owusu-Agyei, S., Lartey, T., Danso, A., Adomako, S., Amankwah-Amoah, J.: CEOs' reputation, quality management and environmental innovation: the roles of stakeholder pressure and resource commitment. Bus. Strateg. Environ. **29**(6), 2310–2323 (2020)
5. Medvedev, V.: Sustainable Development of Society: Models, Strategy. Moscow Academy (2011)
6. Barykin, S., Borovkov, A., Rozhdestvenskiy, O., Tarshin, A., Yadykin, V.: Staff competence and training for digital industry. In: IOP Conference Series Materials Science and Engineering, Vol. 940, No. 1 (2020)
7. Di Berardino, D., Corsi, C.: A quality evaluation approach to disclosing third mission activities and intellectual capital in Italian universities. J. Intellect. Cap. (2018)
8. Degtereva, V., Ivanov, M., Barabanov, A.: Corporate social responsibility as a trend of innovative development in Russia (2022)
9. Estdale J., Georgiadou, E.: Applying the ISO/IEC 25010 quality models to software product. In: 25th European Conference of Systems, Software and Services Process Improvement, pp. 492–503. Springer, Spain (2018)
10. Lyamin, B., Chernikova, A.: Methodology for managing environmental innovation projects at the enterprise. In: E3S Web of Conferences, EDP Sciences, 7007 (2021)
11. Perez-Castillo, R., Carretero, A.G., Caballero, I., Rodriguez, M., Piattini, M., Mate, A., Kim, S., Lee, D.: DAQUA-MASS: an ISO 8000–61 based data quality management methodology for sensor data. Sensors **18**, 9 (2018)
12. Rafique, I., Lew, P., Qanber, A.M., Li, Z.: Information quality evaluation framework: extending ISO 25012 data quality model. Int. J. Comput. Inf. Eng. **6**(5), 568–573 (2012)
13. Kachalov, R.M., Sleptsova, Y.A.: Conceptual model of sustainable development management process. J. Econ. Regul. **12**, 68–84 (2021)
14. Rogova, M.V.: Concept of a mechanism for managing sustainable development of an enterprise. Proc. BSTU Ser. 5 Econ. Manage. **1**(196), 230–234 (2017)
15. Gusev, I.S.: Scientific and methodological support for multi-purpose management of sustainable development of an industrial enterprise. In: Formation of a New Economy and Cluster Initiatives: Theory and Practice, pp. 236–254 (2016)
16. Borgardt, E.A.: Strategic management of sustainable development of an enterprise. Russ. J. Econ. Law **1**(25), 55–61 (2013)
17. Yashin, N.S., Andreeva, T.A.: "Hoshin Kanri" methodology in the strategic management of an oil and gas corporation. Bull. Saratov State Soc. Econ. Univ. **5**(49), 116–122 (2013)
18. Malebnova, S., Tarasov, R.: Process management of quality management systems as part of the formation of a competitive strategy for enterprises in the construction industry. Educ. Sci. Modern World Innov. **1**, 112–121 (2018)
19. Demidovets, V.P.: Formation of a lean production strategy in the forestry complex. Proc. BSTU Ser. 5 Econ. Manage. **7**(189), 198–202 (2016)

Printed in the United States
by Baker & Taylor Publisher Services